# Calculus

## An Active Approach with Projects

The Ithaca College Calculus Group

Stephen Hilbert
John Maceli
Eric Robinson
Diane Driscoll Schwartz
Stan Seltzer

JOHN WILEY & SONS, INC.

New York • Chichester • Brisbane • Toronto • Singapore

*For Alisa, Alison, Elizabeth, Ingrid, J.J., Margaret, Matt, Mike, Monica, Nancy, Peter, Rebecca, Steve, and Sue.*



ISBN 0-471-00316-6

Printed in the United States of America

10 9 8 7 6 5 4 3 2

# Preface

*Calculus: An Active Approach with Projects* is a collection of materials for first-year calculus developed and tested at Ithaca College. It is not a complete textbook, but a complementary volume that can be used successfully in conjunction with any textbook.

We view calculus as a unified subject rather than a linearly ordered sequence of topics and believe that this view should be conveyed to students from the outset of their studies. We have found that students who are actively involved in class are more likely to succeed than those who are passive note-takers. The materials in *Calculus: An Active Approach with Projects* were designed to bring these ideas to life in the classroom.

There are two major sections in the book. The first section contains *activities* that can be done in class or as homework. The second section contains *projects* for the students to work on (usually in teams) outside the classroom.

## Activities

The class activities are designed to accustom students to active participation in the course and to introduce some of the material and methods we have identified as important. Activities involve the students in their own learning. Students who have used activities regularly in class tend to make better comments and ask more significant questions about course material than students in more traditional classes. The key to this difference is active participation. This kind of student involvement fosters understanding and retention of course material.

The activities in *Calculus: An Active Approach with Projects* have several purposes. Through the activities, students participate in the development of many of the central ideas of calculus. Activities introduce new calculus topics, often in a guided discovery format. This reduces the amount of formal presentation that the instructor must do. The activities are also an excellent vehicle for promoting cooperative learning in the classroom.

By doing activities, students learn modeling and how to use the top-down approach to solve problems—both are useful for successful completion of projects, and help them solve shorter problems as well. Students who have experienced activities and projects do not regard "word problems" with dismay.

Activities also help students learn how to draw and interpret graphs—a key element in learning ways to represent functions that are not necessarily given as formulas. Finally, the first set of activities provides an overview of most of first-semester calculus. Doing a number of these activities early in the course helps students see the unity of the subject.

## Projects

Projects serve to reinforce material already presented, motivate concepts, or introduce topics that might not otherwise be covered. Projects bring out both the relevance and the unity of calculus. Most of the projects are set in "realistic" situations. Most also involve more than one calculus topic, often combining topics in unexpected ways. Students need to broaden their perspective and synthesize ideas in order to complete these projects successfully.

We have students work on the projects in teams of three or four, submitting a single report, although many of the projects have parts that are completed by individual students. To complete a project, each team needs to submit a well written report of its solution. Writing about mathematics may be a new experience for many students, but it is a valuable one. To describe the solution of a significant problem precisely in words requires a deeper understanding than most students gain from just solving many problems that are based on examples found in their notes or textbooks. Most of the projects require about two weeks to complete. In a typical semester course, we have our students do three or four projects.

Many of the projects have "open-ended" parts—that is, parts for which there is not a unique correct answer or approach. These questions encourage the students to brainstorm with their teams and to view mathematics as a subject with creative elements.

A significant benefit of this project-oriented approach is that students learn to solve non-trivial, multi-step problems. Working on the (shorter) activities in a guided classroom environment helps them succeed on projects.

## Spiral Approach

The organization and content of the early activities in this book are based on the *spiral approach.* We have found this to be a particularly effective way to teach calculus. Our goal is to present the main ideas of the course early, without getting mired in detail, so that the students will see calculus as a unified subject. The emphasis at this stage is on concepts and relationships, not on technical details.

We use the *calculus of graphs* for this purpose. That is, representing the functions involved almost exclusively in graphical form, and using the familiar ideas of velocity and distance as examples, we examine basic ideas from both differential and integral calculus. Within days, the students have some basic understanding about rates and slopes, concavity, and integration (in the context of obtaining a distance graph when given the corresponding velocity graph). During the rest of the course, students encounter these ideas again and again, each time picking up more of the technical and computational details.

## "New" Calculus

Efforts to revise the way calculus is taught have focused on a number of different issues. The materials presented in *Calculus: An Active Approach with Projects* are designed to empower the student to take an active role in her or his own learning. We emphasize the role of calculus as a tool for understanding the world and hence focus on modeling as a central theme. We also emphasize the notion of function and are careful to show that functions can be represented in many different ways: as graphs, as tables of values, as algebraic expressions, as verbal descriptions, as physical relationships, and as theoretical models. These materials are designed to enable any instructor to incorporate these ideas and approaches into their calculus course.

## Technology

Some of the activities presuppose the use of either a computer or a graphing calculator. Some of the projects are greatly simplified if some computational device is available to help with the calculations and graphs. The *Instructor's Guide* lists what, if anything, is needed for any particular activity or project. We do not prescribe any particular choice of technology, however. We have always described our materials as *technology independent.* This means that most of the activities and projects require no technology at all, and the few that do require a computer or graphing calculator are presented in such a way that the instructor using the material can choose whatever implementation is available.

## The Instructor's Guide

Teaching a calculus course using "new" materials usually requires some modification of one's teaching style and some reorganization of topics. As an aid in this adaptation, we have made available an *Instructor's Guide* to accompany this book. There you will find annotated versions of the activities and projects, including time estimates, background required, teaching suggestions, and topics covered. We have also provided sample curricula, sample test questions, and a set of answers to frequently asked questions about the materials in the book.

## Acknowledgments

We wish to thank a number of people whose advice and encouragement have been invaluable to us during this project.

Paul Glenn of Catholic University was an original member of our group and a valuable participant during the initial phase of our project.

Professors William Lucas of Claremont Graduate School and Gil Strang of M.I.T. have contributed invaluable support and advice since we started developing these materials. Professor Tom Tucker of Colgate University has been a member of our advisory board since 1988.

Our colleagues on the faculty of Ithaca College have been generous with time and help. We especially want to acknowledge useful discussion and class testing of material by Jim Conklin, John Rosenthal, and Martin Sternstein of the Department of Mathematics and Computer Science. We also want to thank all our students and the participants in our workshops for helping us to correct flaws and clarify earlier versions of these materials.

Spud Bradley, formerly of the National Science Foundation, and Paul Hamill of Ithaca College provided both help and expertise when we were looking for funding. Finally, we wish to thank the National Science Foundation and Ithaca College for their support of this project.

# To the Student

This is a book of activities and projects for calculus. It is designed to help you to understand the basic concepts of calculus and to become a good problem solver.

To use these materials successfully, you should approach them with an open mind and a lot of optimism. The activities are relatively short calculus problems or explorations. Most of them will not look like problems you have seen before. All of them are problems on which almost any first-year calculus student can make significant progress.

Working through the activities should help you understand the basic concepts of calculus and how calculus serves as a tool for understanding the world. The activities will also help you learn ways to approach unfamiliar problems and to make progress in solving them.

The projects are larger problems which will require considerable effort to solve. Many of the projects involve several different calculus ideas, so be prepared to draw on all your mathematical background. Many are also "open-ended" problems. That means that there isn't necessarily just one correct solution. Your ideas for these problems will need to be supported by well thought out explanations that will be convincing to others.

Your instructor will decide just how these materials will be used in your particular course. We hope this book will help make calculus interesting, challenging, and meaningful to you.

# Contents

# Part I

# Activities

# Chapter 1

# Graphical Calculus

## Graphical Calculus and Modeling

The first chapter of this book consists of problems and activities designed to introduce you to many of the important ideas of first-year calculus in a way that encourages you to visualize the objects and actions you are studying. We call this approach the *calculus of graphs.*

We see graphs not only in mathematics, but also in the physical sciences, the social sciences, even the daily newspaper. Graphs are a way of comprehending the world. Graphs give us a way to visualize an ongoing process as a whole. That is, a graph can contain the whole past history of a process, and a prediction of its future progress, in a way that can be comprehended quickly. In short, the usefulness of graphs illustrates the truth of the old saying, "One picture is worth a thousand words."

In many of the activities you will be asked to work with phenomena or functions that have been represented only as graphs, and to sketch graphs to represent real world problems you are studying. In many cases, you will not have a formula that corresponds to the graph—the graph itself contains all the information for the problem. This approach will enable you to see the big picture of calculus, while temporarily postponing many of the technical details.

At the same time, you will get an introduction to *modeling.* Modeling is the process of representing real world problems in mathematical terms so that the methods of mathematics can be used to gain understanding of the original problem. When you sketch a graph based on your observation of some action or from reading a verbal description, you are constructing a mathematical model. The effective use of mathematical models has been responsible for much progress in the physical and social sciences.

The approach to the learning of calculus that we take throughout this book emphasizes both the calculus of graphs to gain understanding and insights, and modeling to help you become a good problem solver.

## Chalk toss

Calculus is a study of changes. One form of change is the change in the position of something that is moving. For example, if your instructor throws a piece of chalk into the air and then catches it, the distance of the chalk from the floor changes.

We can record the chalk's position relative to the floor on a graph. The graph will capture the information about how the position changes over time.

1. Watch as your instructor tosses the chalk, and record what you see on a graph.

Exchange papers with the student next to you.

2. Study the graph you just received. In one or two sentences, describe the motion of the chalk that *the graph* describes (even if that does not correspond to what you observed). Be sure to include in your description the answers to the questions:

   "How high did the chalk go?"

   "How long was the chalk in the air?"

3. Briefly discuss both graphs and both verbal descriptions with your neighbor. Decide together which graph you prefer as a solution to the original problem, or what a solution that is better than either one would look like.

4. Watch as your instructor repeats the chalk toss. Based on your previous experience, and your discussion with your neighbor, sketch a new graph.

Watch as your instructor tosses the chalk several more times. After each toss, graph what you see, and compare your new graph with the previous ones.

Your instructor may ask you to discuss your work with another student, and synthesize ideas.

## Classroom walk

Your instructor will walk across the front of the classroom. He or she will designate a "starting line" in the front of the room, so there will be a reference point for the trip.

1. You are to graph the displacement, $f$, of your instructor from the starting line, as a function of time, $t$.

2. Graph the second trip.

3. Graph the third trip.

**Problems**

1. Biking to school
2. Raising a flag

## Biking to school[1]

Terry usually rides a bicycle to school. Below are four graphs and three explanations. Match each explanation with a graph, and write an explanation for the remaining graph.

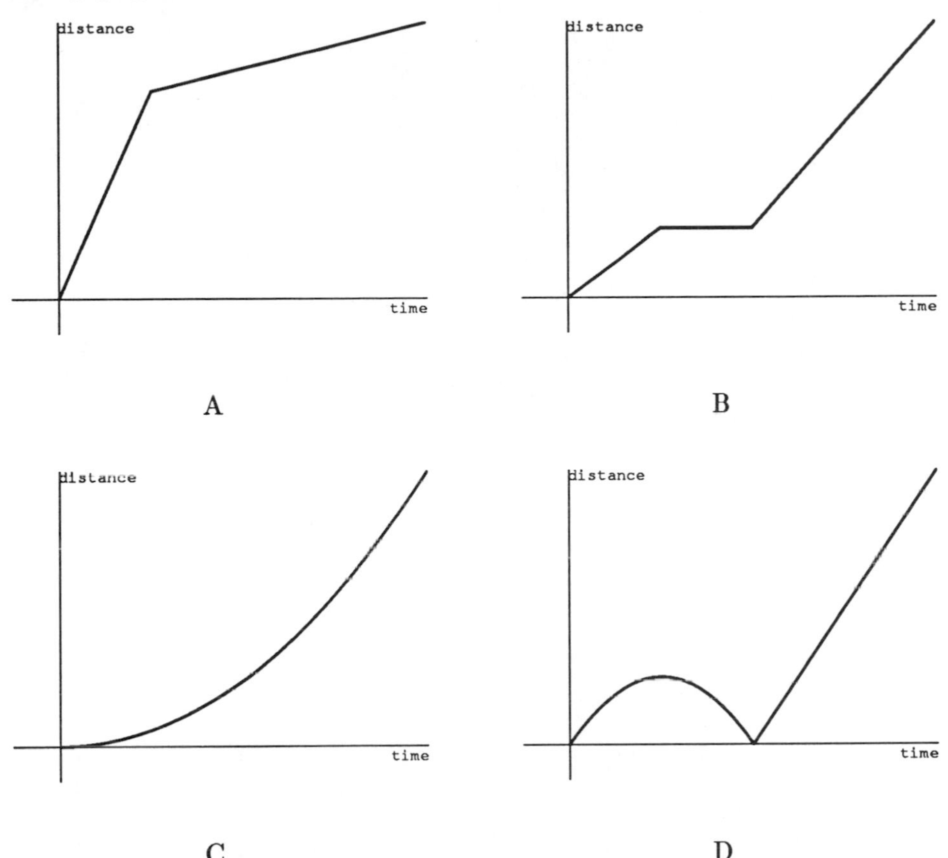


1. "I had just left home when I realized we have gym today, and I had forgotten my gym clothes. So I went back home and then I had to hurry to be on time."
2. "I always start off very calmly. After a while I speed up, because I don't like to be late."
3. "I went on my motor bike this morning, very quickly. After a while, I ran out of gas. I had to walk the rest of the way and was just on time."
4.

[1] Adapted from Neil Davidson, ed., *Cooperative Learning in Mathematics: A Handbook for Teachers*, Addison-Wesley, 1990.

## Raising a flag

1. Sketch the height of a flag as a function of time as it is being raised.

2. Sketch the height of a flag as a function of time as it is being lowered.

## Library trip

In this activity we will describe a situation verbally and ask you to try to construct a graph that might correspond to it.

> Josh had arrived a little early for his calculus class when he realized that he had left his notebook in the library. Not wanting to miss the beginning of class, he hurried directly to the library, picked up his notebook, and returned just as quickly to the classroom. The library is 500 meters directly across the quad from the classroom, so you can assume he walked back and forth in a straight line. The entire trip took six minutes.

1. Construct a graph describing Josh's trip to the library and back. The independent variable is time, $t$, and the dependent variable is the distance, $f$, between Josh and the classroom.

2. Now let's embellish the story a little:

   > On the way to the library Josh met Pat and stopped to talk for three minutes. Then he had to move even faster for the rest of the trip. It took him a total of eight minutes to make the entire trip.

   As before, record the trip on a graph.

3. Now that we have seen how motion can be recorded on a graph, let's look at some of the graphs you have drawn, and see how important information is shown on the graph.

   Let's look first at a typical graph for question 2—Josh's interrupted trip to the library.

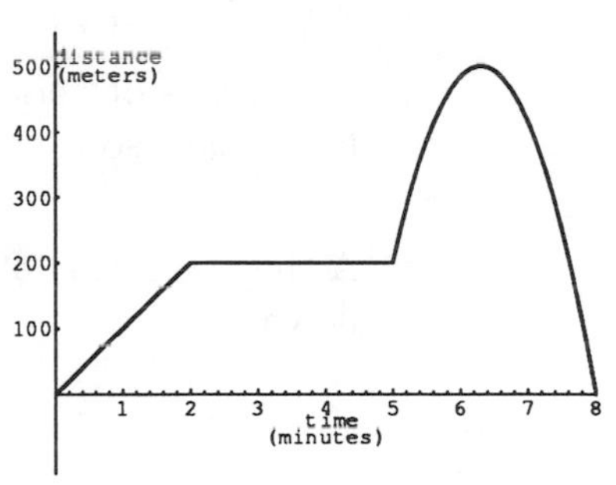


Library trip

Label on the graph the locations of the following (label the graph with the letter of the corresponding question):

(a) The period of time that corresponds to Josh's conversation with Pat.

(b) The time or period of time that corresponds to Josh's presence in the library.

(c) A period of time when the distance between Josh and the classroom is increasing.

(d) A period of time when the distance between Josh and the classroom is decreasing.

(e) A time or period of time when the distance between Josh and the classroom is a maximum.

(f) A time or period of time when Josh is not moving at all.

An important concept associated with motion is that of velocity. Velocity represents both the speed at which something is moving and the direction in which it is moving. For this to make any sense, we have to agree in advance about what we will regard as the "positive direction." The opposite direction is then the "negative direction."

For this problem, suppose we agree that an arrow pointing from the classroom toward the library is pointing in the positive direction, and an arrow pointing in the opposite direction is pointing in the negative direction. Thus, when Josh is moving away from the classroom, his velocity is positive, and when he is moving toward the classroom, his velocity is negative. Aside from these considerations, velocity is just speed. That is, Josh's speed is the absolute value of his velocity.

4. Now try to label these additional areas on the graph:

   (a) A period of time when the velocity is positive.

   (b) A period of time when the velocity is negative.

   (c) All the times when the velocity is zero.

   (d) Two times or periods of time, $4d_1$ and $4d_2$, when the velocity is positive, but chosen so that the velocity at $4d_2$ is greater than the velocity at $4d_1$.

   (e) A time or period of time when the velocity is positive, but Josh is slowing down.

## Airplane flight with constant velocity

### Part A

A plane flies over Ithaca on its way to Los Angeles at 1 o'clock in the afternoon. The plane's velocity is 500 mph, and for the next four hours the plane flies at a constant velocity of 500 mph. The plane flies at the same altitude for the entire time.

1. Sketch a graph of the velocity of the plane versus time.

Since the velocity at 3:30 is 500 mph, your velocity graph should have a point with coordinates $(3.5, 500)$. Remember the first coordinate in the coordinate pair indicates the time. Since you were not told what the velocity is at 12:15, there is no point on the graph whose first coordinate is 12.25.

2. Sketch a graph of the *distance* of the plane from the place where it passed over Ithaca versus time.

3. Compute a table of values of the distance at 1, at 2, at 3, at 4 and at 5 o'clock to check your graph in 2. If you were not able to sketch the graph, use the table of values to locate points on the graph and then go back and sketch the graph.

| Time | Distance |
|---|---|
| 1 | |
| 2 | |
| 3 | |
| 4 | |
| 5 | |

4. To make sure you understand this example draw velocity graphs of a plane that had a velocity of 600 mph and a plane that traveled at a velocity of 400 mph. Then draw the distance graphs for these planes on the same set of axes as you drew the distance graph of the original plane.

5. Look at the distance graphs and try to relate the distance graphs to the velocity graphs.

When you learned about straight lines you learned the concept of the slope of a line. If you are graphing a variable $y$ on the vertical axis and a variable $x$ on the horizontal axis and $(x_1, y_1)$ and $(x_2, y_2)$ are the coordinates of two points, then the *slope* of the line segment between the two points is $(y_2 - y_1)/(x_2 - x_1)$.

Another very important way to think of the slope is that it gives the *rate of change of the quantity measured on the vertical axis with respect to the quantity measured on the horizontal axis.* Let's work through that phrase in detail for the example we just worked out to make sure we understand what the phrase means. We plotted distance from where the plane passed over Ithaca on the vertical axis and time on the horizontal axis. Two points on the distance graph were $(3, 1000)$ and $(5, 2000)$. Computing the slope between these two points gives you $1000/2 = 500$.

Keep in mind that the vertical axis is measured in miles and the horizontal axis in hours, so we should interpret the slope as a way of telling us that the distance must increase by 1000 miles when the time increases by 2 hours or by 500 miles whenever the time increases by 1 hour. This is written as 500 miles per hour, which is exactly the velocity of the plane.

So we can think of *velocity* as *the rate of change of distance with respect to time.* This is indicated by the units used to measure velocity such as miles per hour, centimeters per second, etc.

Geometrically, a slope of 500 means that to stay on the graph we move up 500 units each time we move 1 unit to the right.

6. Find the slope of the distance graph if the plane had a constant velocity of 400 miles per hour. Be sure to include units of measure.

7. Based on this activity, we have seen that *if the distance graph is a straight line then the* ______________ *of the straight line is the velocity.*

## Part B

At the start of Part A, you were able to draw the graph of the distance travelled by looking at the velocity graph and reasoning about the physical quantities involved. Let's look for a way to use this same reasoning to represent distance travelled *directly on the velocity graph.*

1. Consider any time period, say for example the first half hour (from 1:00 to 1:30). How do you calculate the total distance travelled? It is the velocity multiplied by the elapsed time. That is, $d = v \times t$. For the plane travelling at 500 miles per hour, calculate the distance travelled:

   (a) in the first half-hour.

   (b) in the first hour.

   (c) in the first two hours.

   (d) in the first three hours.

In this example, the velocity graph is a horizontal line, and the magnitude of the velocity at any time is the vertical length measured from the $t$-axis to the velocity graph, always 500 mph in this case. The elapsed time is a length measured along the horizontal axis. In this case it is the length of the part of the horizontal axis between $t = 1$ and $t = 1.5$, that is, 0.5.

So the distance travelled, 500 mph $\times$ 0.5 hr, or 250 miles (see 1a) is also the *area* of the rectangle shaded at the right.

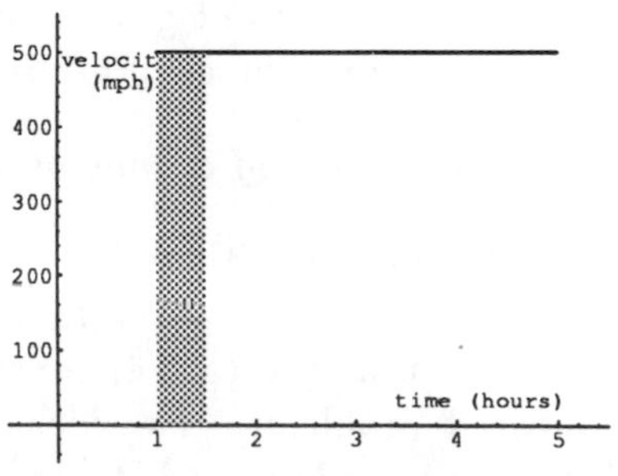


First half hour

We could calculate the distance travelled in any other time period in a similar way. In each case, the distance travelled is also the area of a corresponding rectangle.

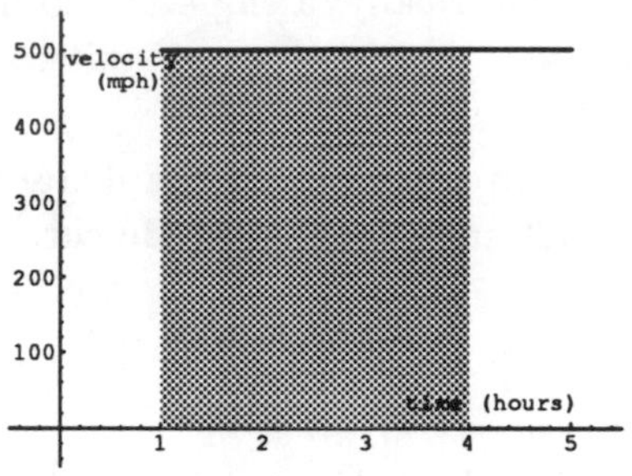


First three hours

2. Calculate the area of the rectangle above. Compare your answer with 1d.

3. Draw a rectangle whose area corresponds to:

   (a) the distance you computed in 1b.

   (b) the distance you computed in 1c.

4. For this example (constant velocity and linear distance), at least, we seem to have found the following pair of principles:

   *If you have a graph of distance travelled, velocity at some specific time is the __________ of the distance graph (at the appropriate point).*

   *If you have a graph of the velocity, distance travelled during some time period is the __________ __________ the velocity graph (over the appropriate interval).*

We have also discovered another relationship. We have found a correspondence or pairing of two graphs. Namely:

*If you have a horizontal velocity graph, then the distance graph is linear.*

*If the distance graph is linear, then the velocity graph is horizontal.*

This idea of pairs of graphs will reappear throughout the course. Finally, to finish summarizing the general principles we have developed:

*Velocity is the rate of change of distance with respect to time.*

In real life (even with cruise control) it would be difficult to fly a plane or drive a car at *exactly* the same speed for a long period of time. However we will use simple examples such as the case of constant velocity to analyze more realistic and more complicated problems. The idea of using simple problems to attack complicated problems is a fundamental strategy in mathematics.

**Problems**

1. For the plane in the preceding section that travelled at a constant velocity of 500 mph, express the velocity, $v$, as a function of time, $t$. That is, write a formula for $v$ in terms of $t$.

   Also express the distance travelled, $f$, as a function of $t$. That is, find $f(t)$.

   Observe how the principle "velocity = slope of distance" looks in terms of the formulas.

   Repeat the problem for the other two velocities, 400 mph and 600 mph.

2. Suppose the velocity is 500 mph from 1 to 3 and 400 mph from 3 to 5. Construct the distance graph using the slope-velocity relationship (your graph will look like two line segments joined together), and verify that distance = area.

## Projected image

### Part A

1. Watch as your instructor moves the overhead projector. Describe in your own words what happens.
2. Now collect some experimental data. Measure the distance of the projector from the wall and the image size at three different points. Record your experimental data in the table below.

| Distance from wall | Size of image |
|---|---|
| | |
| | |
| | |

3. Now, as a group, sketch what you believe is a reasonable graph of size of image versus distance from the wall. Be sure to label your axes clearly and include a scale and units.
4. Write a verbal description of what your graph says happened as the projector was moved.
5. Compare the verbal description you just wrote with the description you wrote before collecting the data and drawing the graph.
6. Now choose a distance that you did not measure experimentally, and predict what the image size would be at that distance. Measure and check your prediction.

### Part B

Below is the experimental data collected in one of our classrooms. As with the data you collected, some of our measurements may be in error. We decided to measure "size" as the height of the projected image. We have transferred the points to a graph.

| Distance from wall | Size of image |
|---|---|
| 18 | 13 |
| 24 | 17 |
| 48 | 29.5 |

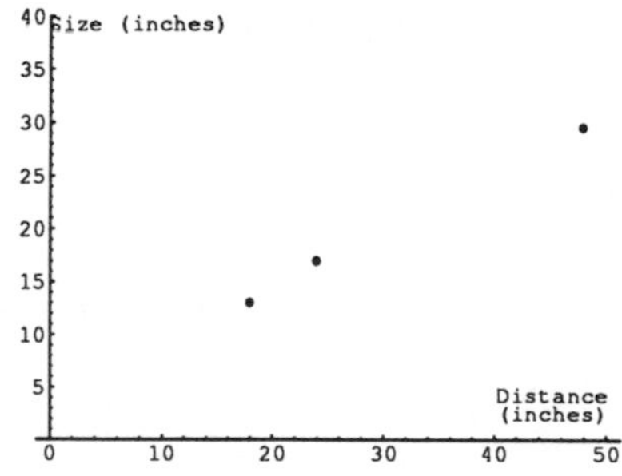


Projector data

As you probably observed when you made your own measurements, errors in making the measurements have almost certainly occurred, so even the little data we have is not necessarily accurate. And the three points on the graph, of course, give a very incomplete picture of the relationship between distance and image size. For instance, how big is the image when the projector is 42 inches from the wall? Or at any other distance, $x$, from the wall?

1. Based on the three data points given above, predict the size of the image when the projector is 42 inches from the wall.

2. If we have only this experimental data, what kinds of graphs are reasonable models based on the data? *The key question to ask is: what happens between the data points?* There are lots of ways to fill in the graph. Several possibilities are given on the next page. For each graph given:

   (a) What size image does the graph predict when the projector is 42 inches from the wall?

   (b) Describe in your own words why it is or is not a reasonable graph to use to represent the data and the experiment.

   (c) Choose the graph that you think best describes the size of the image as a function of the distance of the projector from the wall.

3. For as many of the graphs as you can, find a formula for the size of the image in terms of the distance from the wall.

4. Suppose the projector is 24 inches from the wall. For each graph, predict what would happen if the projector is moved a short distance

   (a) closer to the wall.

   (b) further from the wall.

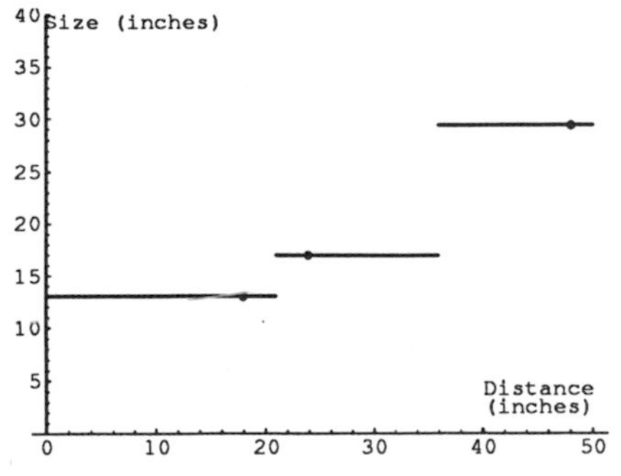


A

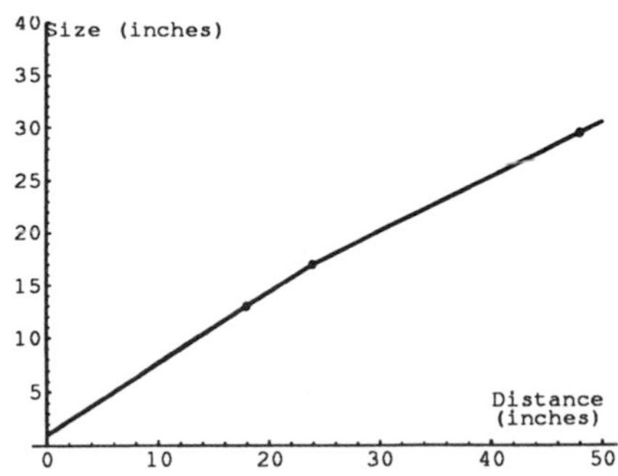


B

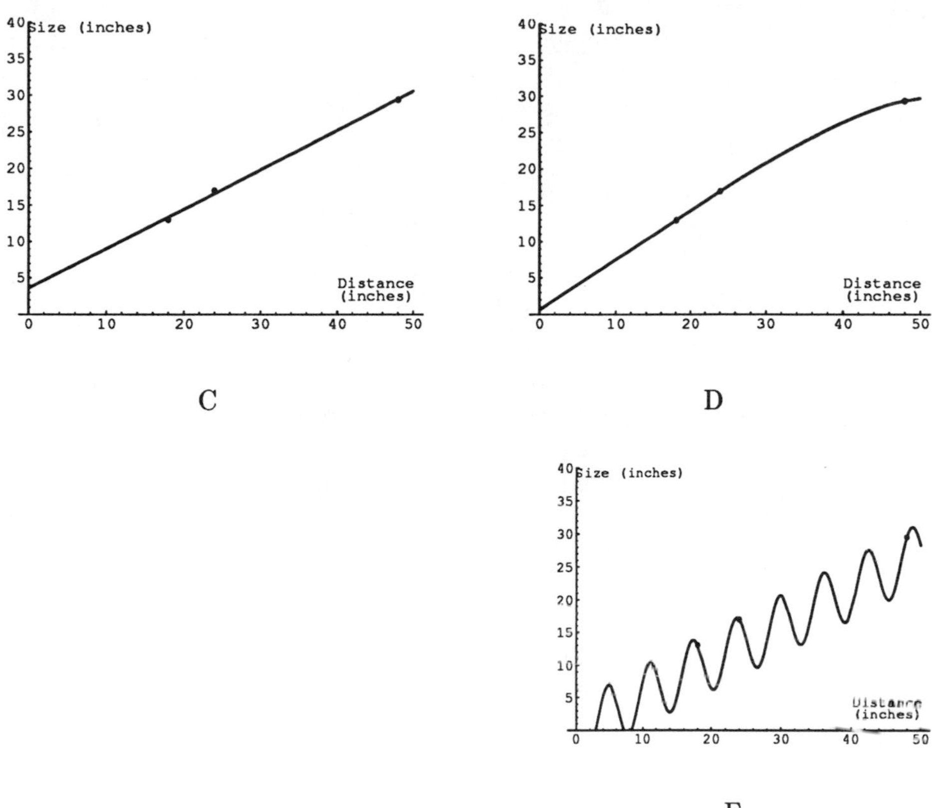


**Problems**

1. What does each graph above tell you about the size of the image when the projector is 36 inches from the wall?

2. Describe in a few sentences what each graph says about how the size of the image changes as the projector is moved from the wall.

## A formula for a piecewise-linear graph: Top-down analysis

The second graph from the projected image activity is shown at the right.

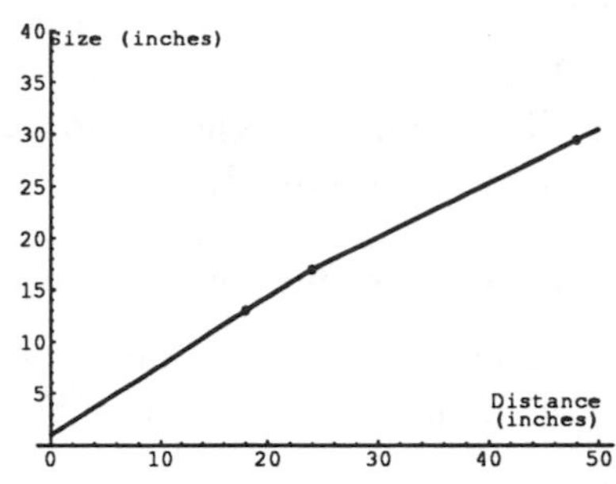


Projector image

This is a piecewise-linear graph; that is, it is obtained by "pasting together" line segments. In this section we will discuss how to get a formula for this (and any other) piecewise-linear graph. Let us look in some detail at how this is done.

Our solution to this problem uses a top-down approach. That is, we will determine what big steps are needed to solve the problem, then reduce each big step to smaller steps. We repeat this process, breaking each smaller step into even smaller ones until we recognize the smallest pieces as problems we know how to solve. We then reassemble the solutions to the small pieces successively until we have a solution to the original large problem.

Let's see how this works for the problem of writing a formula for the piecewise-linear graph.

Level I (top level)

The formula we want describes two line segments; call them $L_1$ and $L_2$. The formula for line $L_1$ applies only between the first and second data points, and the formula for $L_2$ applies only between the second and third data points. So we can find the final answer if we can find the answers to two smaller problems:

I.A. Find a formula for line $L_1$ when $x \leq 24$, and

I.B. Find a formula for line $L_2$ when $x \geq 24$.

So the first step in the top-down analysis of the problem (and the last step we do when we are reassembling the pieces) is:

I. Write the formula for the piecewise-linear graph as:

$$y = \begin{cases} \text{formula for line } L_1 & \text{when } x \leq 24 \\ \text{formula for line } L_2 & \text{when } x \geq 24. \end{cases}$$

Level II (smaller problems we need to solve in order to complete level I)

It should be pretty clear that in order to complete I, we need to be able to do two things:

Write the formula for $L_1$, a line that goes through the two points $(18, 13)$ and $(24, 17)$.

Write the formula for $L_2$, a line that goes through the two points $(24, 17)$ and $(48, 29.5)$.

So if we can solve these two problems, then we can solve the original problem. This leads us to level III.

Level III (smaller problems we need in order to complete level II)

In order to find the equation of a straight line when we are given two points we can use the point-slope form of the equation for a straight line, which is

$$y - y_1 = m\,(x - x_1).$$

In this formula, $m$ is the slope of the line and $(x_1, y_1)$ are the coordinates of a point on the line. So in order to solve I.A we need to solve three smaller problems:

I.A.1. Find the slope of the line through $(18, 13)$ and $(24, 17)$.

I.A.2. Use one of the two given points and the point-slope formula to find the equation of the line through $(18, 13)$ and $(24, 17)$.

I.A.3. Solve the equation that we found in I.A.2 for $y$ to get the formula for $L_1$.

Similarly, in order to find the equation for $L_2$ we will need to solve the same three problems.

Now we will solve these problems and assemble a solution to the original problem.

First we will solve I.A.1.: the slope is $\frac{17-13}{24-18} = \frac{2}{3}$. Using $(18, 13)$ as the given point gives $y - 13 = \frac{2}{3}(x - 18)$ as the result of I.A.2. Finally, solving for $y$ gives us $y = \frac{2}{3}x - 12 + 13 = \frac{2}{3}x + 1$ as the outcome of I.A.3.

Similarly, the slope is $\frac{29.5-17}{48-24} = \frac{25}{48}$. Using $(48, 29.5)$ as the given point gives $y - 29.5 = \frac{25}{48}(x - 48)$. Finally, solving for $y$ gives us $y = \frac{25}{48}x - 25 + 29.5 = \frac{25}{48}x + \frac{9}{2}$.

Solving the level III problems has given us the solution to the level II problems.

The equation of line $L_1$ is $y = \frac{17-13}{24-18}(x - 18) + 13$ or $y = \frac{2}{3}x + 1$ and the equation of line $L_2$ is $y = \frac{29.5-17}{48-24}(x - 24) + 17$ or $y = \frac{25}{48}x + \frac{9}{2}$.

These two equations are the solutions to the two subproblems at level II.

Returning finally to level I of the original problem (reassembling):

The solution is: the formula for the piecewise-linear graph is

$$f(x) = \begin{cases} \frac{2}{3}x + 1, & x \leq 24 \\ \frac{25}{48}x + \frac{9}{2}, & x \geq 24. \end{cases}$$

We can often summarize the top-down analysis of a problem in outline form. The outline looks a bit like one you would use in organizing an English paper. The outline should really be developed in advance and be used as a guide in the solution of the problem. We have reversed the procedure here to help you appreciate the thinking that goes on in designing the solution.

Here is the outline form of the top-down structure of our solutions.

Problem: Write a formula for the piecewise-linear graph.

**I** Write the formula for the piecewise-linear graph as

$$y = \begin{cases} \text{formula for line } L_1 & \text{when } x \leq 24 \\ \text{formula for line } L_2 & \text{when } x \geq 24. \end{cases}$$

**A** Write the formula for $L_1$, a line that goes through the two points $(18, 13)$ and $(24, 17)$.

1. Find the slope between $(18, 13)$ and $(24, 17)$.
2. Find the point-slope equation of the line between $(18, 13)$ and $(24, 17)$.
3. Solve for $y$ to get the formula for $L_1$.

**B** Write the formula for $L_2$, a line that goes through the two points $(24, 17)$ and $(48, 29.5)$.

1. Find the slope between $(48, 29.5)$ and $(24, 17)$.
2. Find the point-slope equation of the line between $(48, 29.5)$ and $(24, 17)$.
3. Solve for $y$ to get the formula for $L_2$.

## Problems

1. Fit a piecewise-linear graph to the three experimental data points you collected in class. Determine the corresponding formula for the graph.

2. Observe the length of the cafeteria line in the dining hall at three different times. Record line length versus time in a table. Fit a piecewise-linear graph to the data. Include both a sketch of the graph and its formula.

## Water balloon

This is another physical experiment in which we will compare observed data with a mathematical model. Your instructor has a balloon filled with water. The water will be released into a beaker a little at a time. At several points in the process you will measure the circumference of the balloon, and also observe how much water is in the beaker. That is, you will be observing the relationship between the balloon's circumference and the amount of water that has been released.

Record your experimental data in the table below.

| Circumference in ______ (units) | Amount of water released in ______ (units) |
|---|---|
| | |

To predict the volume of water released at circumferences between those you observed you could fit a piecewise-linear or other graph to the data. In fact, you will probably be asked to do that as a homework problem.

Another approach is to *make a mathematical model in the form of a formula directly from the (theoretical) geometric properties of the balloon and the stated description of the problem.*

Before you begin, let us agree to call the circumference of the balloon $C$, and the amount of water released $A$. The formula you derive, therefore, will be of the form

$$A = \text{some formula whose variable is named } C.$$

In fact, to emphasize that the value of $A$ will depend on the value of $C$, we will write $A(C)$ in place of $A$. That is,

$$A(C) = \text{some formula whose variable is named } C.$$

Find a formula for $A(C)$.

Now that we have a formula for the amount of water released, based on geometric considerations, we should compare the values predicted by our formula with those we obtained by actual measurement. That is, compute $A(C)$ for all the values of $C$ that you measured, and compare the values generated by the formula with the amounts of water you measured.

**Problems**

1. Fit a piecewise-linear function to the experimental data points you collected in class. Determine the corresponding formula for the graph. If you have graphing software available, compare the graph of the formula you derived in class with the piecewise-linear one above.

2. Derive a formula for the projected image model, from theory. (See the "Projected image" activity.)

## Graphical estimation of slope

In this activity, we will examine the concept of "slope" for graphs that are not straight lines. The idea is the following: if you consider only a very small portion of a curve, that curve might appear to be a straight line. As an analogy, assuming that the earth is a sphere (it's not, but it's close), you could walk along the equator and feel that you were walking along a straight line, even though an observer in space would realize that the equator is curved.

The same is true of certain non-linear graphs or curves. Select a point on the graph on which to focus your attention. If you look at a small enough piece of the graph near your point, it looks like a straight line. That apparent straight line has a slope. This slope is a good approximation of what we call the slope of the graph at the point where you are focusing attention. (Some graphs never start to look straight no matter how closely you look. We'll discuss that problem in the second part of the activity.)

### Part A

1. We will begin to examine these ideas by looking at the function $f(x) = 10(x^2 - x^3)$, near the point $(0.5, 1.25)$. First, you should verify that the point $(0.5, 1.25)$ lies on the graph of $f$. (How?)

2. Now, using graphing software on a computer or a graphing calculator, graph $f$. Be sure that your view includes the point $(0.5, 1.25)$ and is large enough to give you a good overall view of the shape of the entire graph.

Now we want to get closer and closer views of this graph, near the point $(0.5, 1.25)$.

Your instructor will tell you how to "zoom in" on a point on the graph, using your software or calculator. In some cases this may be just a matter of selecting a small rectangle on the display, and "blowing up" the picture. In others, you may need to select the $x$ and $y$ limits of the display explicitly.

3. Blow up the display, using a rectangle from about $(0, 0.75)$ (lower left) to $(1, 1.75)$ (upper right). Does the graph appear to be a straight line? If not, blow up the display again. This time use a rectangle from about $(0.4, 1.15)$ to about $(0.6, 1.35)$. Keep zooming in until the graph seems to be a straight line.

4. Once you are satisfied that the graph looks straight from close up, estimate its slope from the display. This requires reading the coordinates of two points from the screen, since we need two points to determine slope. Fortunately, we know one point that is displayed on this part of the graph—it is the point on which we have been focusing attention, $(0.5, 1.25)$.

   Now we need to find a second point on the graph. Your software may have a feature thats helps you find the coordinates of a point on a display. If not, you will have to estimate it as best you can. Record the (approximate) coordinates of the second point.

   Use these two point to calculate the approximate slope of the line.

   When you finish, redisplay the graph, returning to the original domain and range, and continue.

5. Zoom in once again on the graph of $f(x)$, this time focusing on the point $(0, 0)$. Find two points and the slope.

6. The second point we used in computing the slope of the "line" is just a visual estimate. That is, we cannot be sure that it is actually a point on the "line." How can you exploit the fact that this apparent "line" is actually a portion of the original graph in order to get exact coordinates of a second point?

**Part B**

1. Now that you know how to estimate the slope of a function graphically, you are to obtain such estimates for several points on the graph of $g(x) = 2x - 3x^2$.

Fill in the table at the right.

| Focus point | Approximate slope |
|---|---|
| $(-2,-16)$ | |
| $(-1.5,-9.75)$ | |
| $(-1,5)$ | |
| $(-0.5,-1.75)$ | |
| $(0,0)$ | |
| $(0.5,0.25)$ | |
| $(1,-1)$ | |
| $(1.5,-3.75)$ | |
| $(2,-8)$ | |

2. When you have completed your table, observe whether or not there seems to be any pattern. Can you guess a way of determining the slope at the point $(1.1,-1.43)$ without using the computer? How about at the point $(a, 2a-3a^2)$?

**Part C**

1. See what happens for the function $h(x) = |x|$.

   Fill in the table at the right.

| Focus point | Approximate slope |
|---|---|
| (-1, 1) | |
| (-0.5, 0.5) | |
| (0, 0) | |
| (0.5, 0.5) | |
| (1, 1) | |

2. When you have completed your table, observe whether or not there seems to be any pattern. Pay special attention to $(0,0)$. How about at the point $(a,a)$ for $a > 0$? How about at $(a,-a)$ for $a < 0$?

## Slope with rulers

For each of the four graphs below, estimate the slope at a number of points. Can you find a function that gives the slope as a function of $x$?

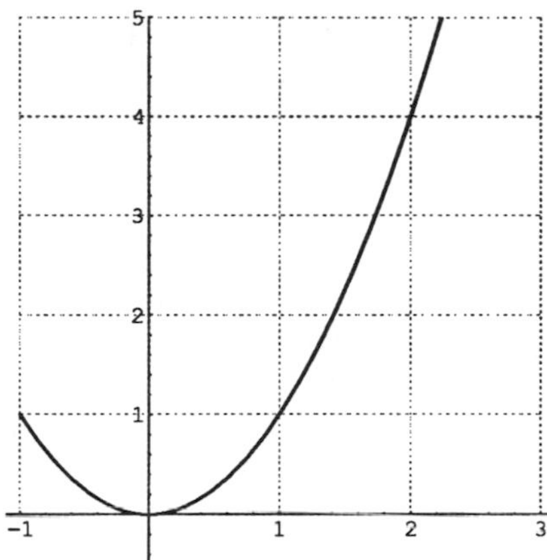


A

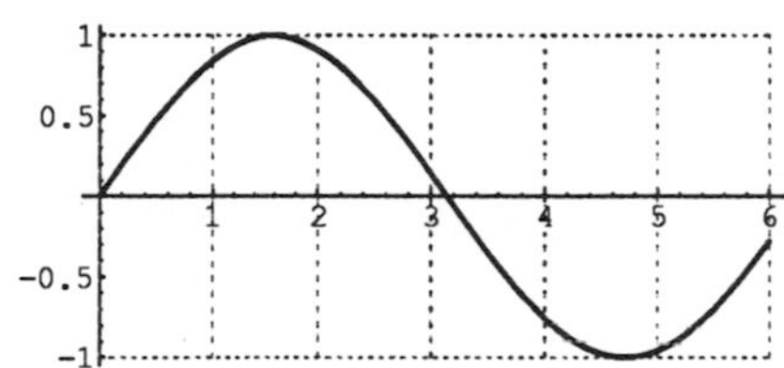


B

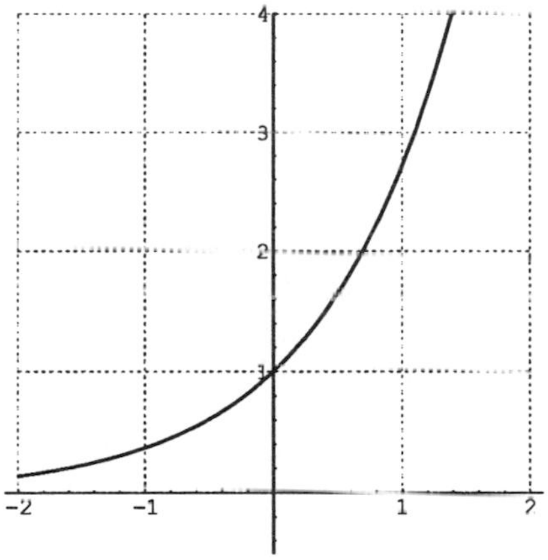

C

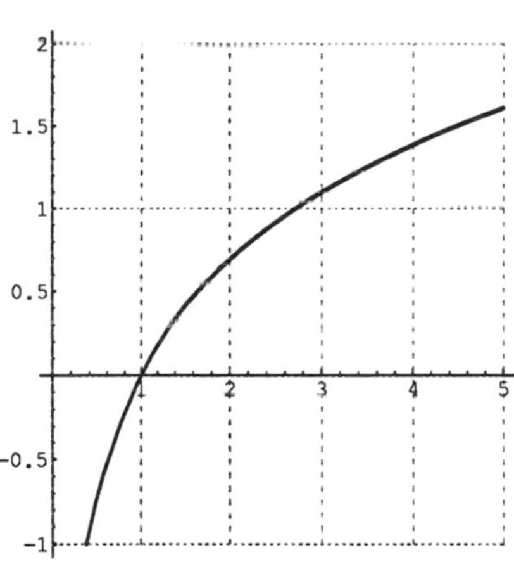


D

## Examining linear velocity

Returning to the relationship between distance and velocity, suppose we have a vehicle that starts at rest (velocity = 0) and speeds up at a constant rate. (By that we mean that the vehicle speeds up in such a way that its velocity graph is linear.) Let's assume the velocity has the formula $v(t) = 2t$, for $0 \leq t \leq 8$. Assume that time is measured in seconds and distance in feet, so velocity is measured in feet per second. Then the velocity graph looks like this.

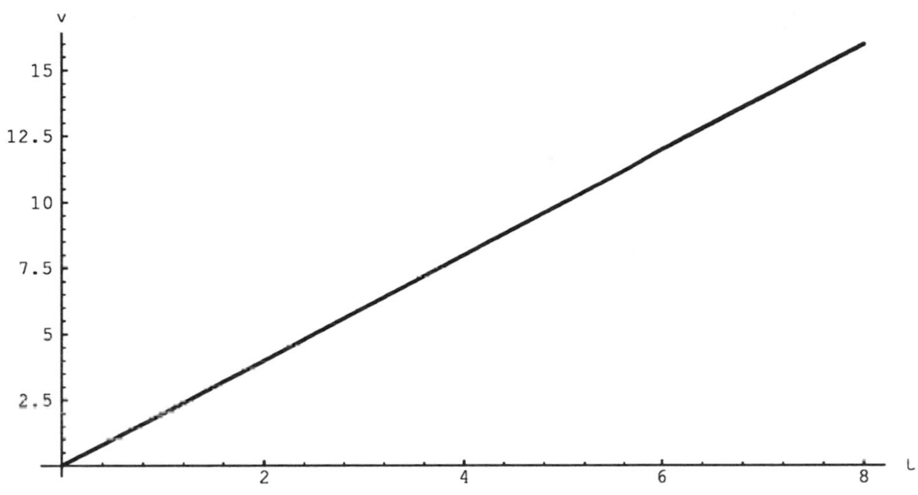


Constant velocity

Can we sketch a reasonable graph of the distance travelled?
Do the two principles

distance travelled = area under the velocity graph
and
velocity = slope of the distance graph

still seem to hold?

### Part A

1. Finding out exactly what the distance graph looks like is no longer such an easy problem. One way we can approach it is by computing some approximations. During the first second, the car speeds up from 0 feet per second (fps) to 2 fps. So we might realistically say that its average speed during the first second is 1 fps. A reasonable approximation of how far it travels in the first second is obtained by pretending it travels at 1 fps for the entire second. The distance, therefore, is approximately 1 fps $\times$ 1 sec, or 1 foot.

Similarly, the average velocity for the second second might reasonably be said to be 3 fps, since the speed at the beginning of the second second is 2 fps and at the end of the second second it is 4 fps. Thus, the approximate distance travelled in the second second is 3 fps × 1 sec, or 3 feet. Fill in the rest of the table at the right.

| Second | Computation |
|---|---|
| 1 | 1 fps × 1 sec = 1 ft |
| 2 | 3 fps × 1 sec = 3 ft |
| 3 | 5 fps × 1 sec = 5 ft |
| 4 | |
| 5 | |
| 6 | |
| 7 | |
| 8 | |

2. The approximate total distance travelled in the first $x$ seconds, using $x = 1, 2, 3, 4$ is tabulated at the right. Fill in the rest of the table.

| $x$ | Approximate total distance travelled in first $x$ seconds |
|---|---|
| 1 | 1 |
| 2 | 4 (i.e., $1 + 3$) |
| 3 | 9 (i.e., $1 + 3 + 5$) |
| 4 | 16 $(1 + 3 + 5 + 7)$ |
| 5 | |
| 6 | |
| 7 | |
| 8 | |

3. Now, we can get a rough sketch of the distance graph by sketching a curve through the points $(0,0)$, $(1,1)$, $(2,4)$, ..., $(8,64)$. The actual points we obtained, however, should give you a good idea of just what curve will represent the exact distance graph! What curve will represent the distance?

4. Now let's see how these calculations are related to area. Look again at the velocity graph. We will compute the area under the graph from $t = 0$ to $t = 1$, from $t = 0$ to $t = 2$, ..., and from $t = 0$ to $t = 8$ first using an approximating technique. Look at the diagram below. For the first second, the area under the graph (which is actually the area of a triangle) can be seen to be equal to that of the sketched rectangle. Similarly for the area for the second second, and so on.

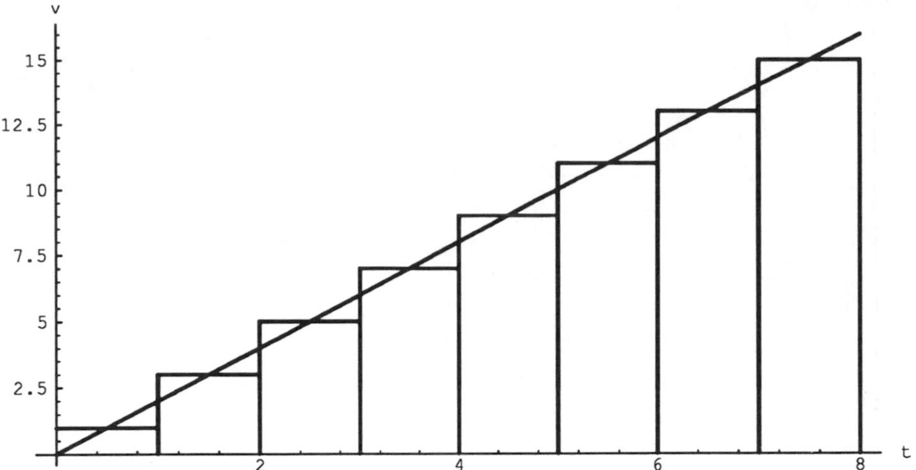


Approximate distance each second

Calculate these areas, along with the total area from $t = 0$ up to and including the current rectangle, in the table at the right.

| $x$ | Total area to the left of $x$ | |
|---|---|---|
| 1 | 1 | |
| 2 | 4 | (i.e., $1 + 3$) |
| 3 | 9 | (i.e., $1 + 3 + 5$) |
| 4 | | |
| 5 | | |
| 6 | | |
| 7 | | |
| 8 | | |

Not only are the values for total area the same as those for our approximation of total distance travelled, the computations are the same too.

Once again, the relationship "distance travelled = area under velocity curve" is seen to hold. Furthermore, we seem to have another pair of related graphs, namely:

*If the velocity graph is linear, then the distance graph is quadratic.*

## Part B

What about the second relationship: velocity = slope of distance graph? In the current problem we have a distance graph whose formula is $f(t) = t^2$.

A carefully drawn graph of $f(t) = t^2$ is given here.

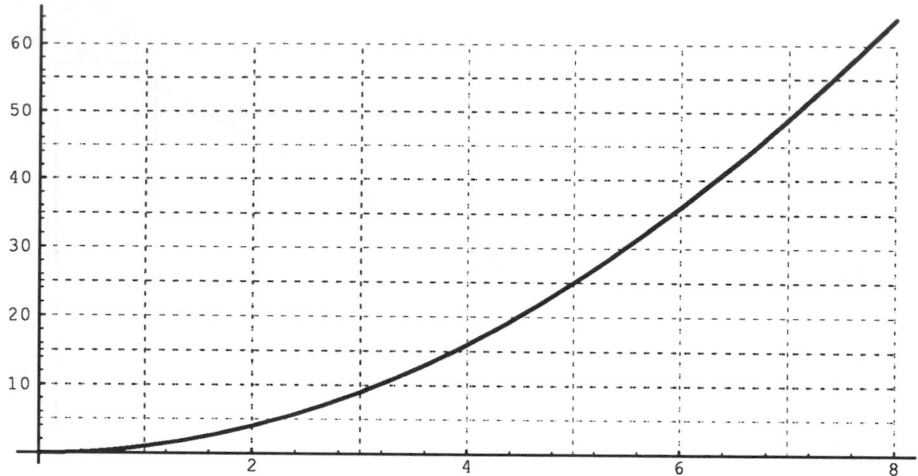


Distance travelled

1. Use the graph and a ruler to estimate the slope of the graph (i.e., the slope of the tangent line to the graph) at each of the values of $t$ that are listed in the table at the right.

| $t$ | Slope of the graph at $t$ |
|---|---|
| 1 | |
| 2 | |
| 3 | |
| 4 | |
| 5 | |
| 6 | |
| 7 | |
| 8 | |

2. Guess a formula for slope, as a function of $t$.

Some helpful ideas for computing the slope with a ruler are:

- Use two points that are far from each other rather than two points that are close to each other in order to minimize the error introduced by reading coordinates from the graph.
- Use points which are on the grid if at all possible.
- Coordinates of points on the graph can be found by using the rule for the function.

3. Does the principle "velocity = slope of distance graph" still seem to be true?

## Problems

1. Suppose $v(t) = 3 - 4t$ for $0 \leq t \leq 4$. Find the distance function. Graph the distance function. Verify that the slope function for the distance graph is $3 - 4t$.

2. Prove by geometry that the area under the curve $f(t) = 2t$ from $t = 0$ to $t = x$ is given by $A(x) = x^2$.

## Given velocity graph, sketch distance graph

We will analyze the general problem of sketching the distance graph if we are given the graph of the velocity. We will approach this using the top-down method.

Problem: Given the graph of the velocity, sketch the graph of the distance traveled.

Level I (top level)

When we sketch a graph we draw a curve. There are many different ways to do this. One common method is to draw and scale the coordinate axes, then plot some points, and finally connect these points. However, if the points are arbitrarily selected we may not obtain the correct graph. For instance, if we were drawing the graph of a piecewise linear function and we plotted points that all corresponded to the same linear piece, we might conclude that the graph was a straight line. So, if we want to solve the problem using this method we need to be sure that we plot points that will not lead us into making a mistake.

This approach breaks the original problem into three smaller problems:

I.A. Scaling the coordinate axes

I.B. Choosing and plotting points on the distance graph

I.C. Connecting the points that we plotted

Level II (smaller problems we need to solve in order to complete level I)

First we will analyze A and B. When we choose the scale for the coordinate axes we want to make sure that we "see" all the meaningful information. For instance if we graphed $y = \sin(x/100)$ using $x = 0, 1, 2, 3,$ and 4 we would not get the correct idea at all. (Plot the five points and see if you would have guessed the correct shape of the graph.) We also need to know the possible values of the second coordinate in order to decide on an appropriate scale for the vertical axis. So we will have to investigate the maxima and minima of the distance.

We know that when the velocity is positive then the distance traveled is increasing, and when the velocity is negative then the distance traveled is decreasing. So the distance will be a local maximum when the velocity changes from positive to negative. Using the same reasoning, the distance will be a local minimum when the velocity changes from negative to positive. So, we want to plot points on the distance graph that correspond to times when the velocity graph changes sign. There are only two ways for the velocity to change sign:

1. The velocity graph passes through the horizontal axis. When this happens the velocity is equal to 0.

2. The velocity is discontinuous at a time with the velocity positive on one side of the point of discontinuity and negative on the other side of the point of discontinuity.

Times when either of these occurs are called *change points of the velocity.*

Since we are given the velocity graph we can easily pick out all the times when the velocity changes sign. The points corresponding to these times should be plotted on the distance graph. In addition, since these points will include all the local maxima and minima they will give us the information we need to choose the scale of our graph.

So at level II we need to perform the following tasks:

II.A. Find all the times when the velocity is 0.

II.B. Find all the times when the velocity is discontinuous.

II.C. Classify each interval between succesive change points as an interval where the distance is increasing or as an interval where the distance is decreasing.

Using just this much information we could sketch the rough shape of the distance graph since we know where the distance increases and decreases. However, if we know whether the graph curves up or down (is concave up or concave down) we can make a more accurate sketch. Remember, when the velocity is increasing then the distance graph will be concave up and when the velocity is decreasing then the distance graph will be concave down.

So investigating where the distance graph is concave up or down will be done at level III.

Level III (simpler tasks to complete level II)

It is easy to read where the velocity is increasing and where the velocity is decreasing from the velocity graph. Times when the velocity changes from increasing to decreasing or from decreasing to increasing correspond to points where the distance graph will change from concave up to concave down or vice versa so we will also plot these points on our distance graph. So at level III our tasks are:

III.A. Find all the times when the velocity changes from increasing to decreasing or from decreasing to increasing. These points are called *inflection points.*

III.B. Decide whether the distance graph is concave up or concave down on every interval determined by successive inflection points.

**Assembling the solutions**

Now we will assemble the results of our simple problems into a solution to the original problem.

First, mark each change point and inflection point on the horizontal (time) axis.

On each interval determined by in this way, the distance graph must be one of the following types of curves

Increasing and concave up

Increasing and concave down

Decreasing and concave up

Decreasing and concave down.

Determine which type the distance graph is on each interval.

Now we know how to sketch the graph between the points chosen in above. So if we plot these points, then we will be able to sketch the distance graph accurately. We have previously discovered that the distance traveled between $t = a$ and $t = b$

is the area under the velocity graph between $t = a$ and $t = b$. Therefore, if we know the distance at any point, we can estimate the distance at any other point by geometry if the velocity graph is straight or by counting blocks on the velocity graph. So, if we know what the distance is at one time, then we can plot all the points we found in the first step on the distance graph.

Next, plot the points on the distance graph corresponding to the times specified in the first step.

Finally, to finish the problem, connect the points plotted in the previous step according to which of the four types of curves describes the distance in that interval.

### Problems

1. For each of the following velocities:

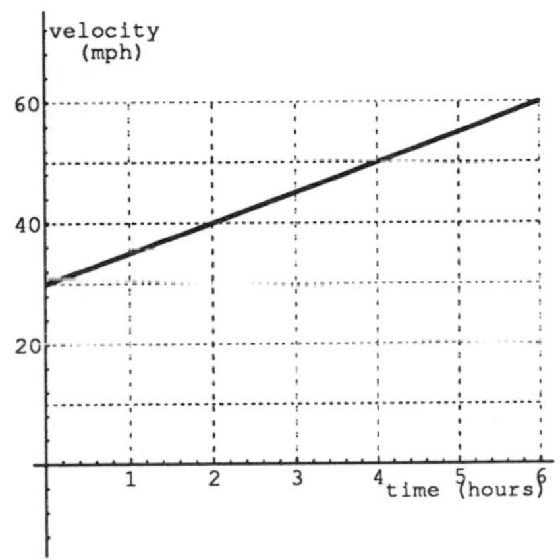


A

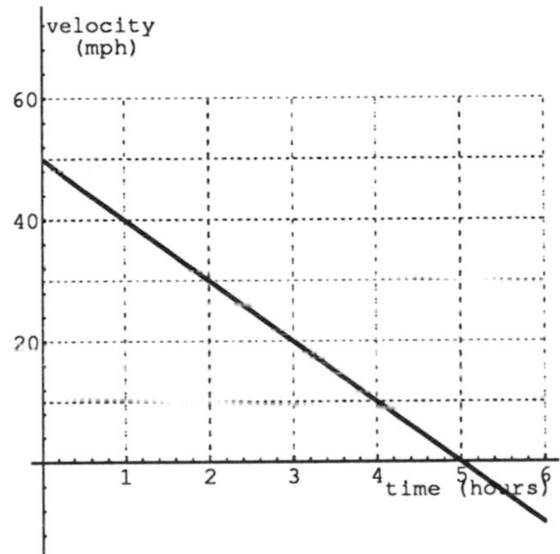


B

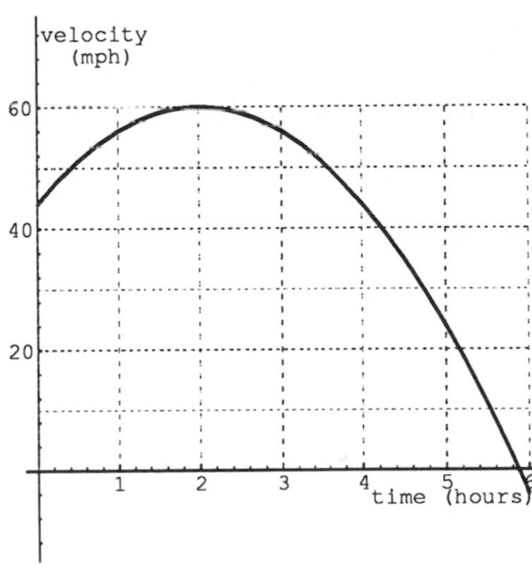


C

(a) Find the distance traveled from $t = 1$ to $t = 3$.

(b) Find the average velocity from $t = 1$ to $t = 3$.

(c) Sketch a constant velocity curve equal to the average velocity from $t = 1$ to $t = 3$.

(d) Find the distance traveled from $t = 1$ to $t = 5$.

(e) Find the average velocity from $t = 1$ to $t = 5$.

(f) Sketch a constant velocity curve equal to the average velocity from $t = 1$ to $t = 5$.

(g) At what time between $t = 0$ and $t = 6$ is the distance from the starting point the greatest?

(h) Find the distance traveled from $t = 1$ to $t = x$.

(i) Sketch the graph of distance traveled versus time.

2. Suppose the temperature at sea level is 200° K and the rate of change of temperature with respect to altitude is $100 - 20t$° K/km at an altitude of $t$ kilometers above sea level.

   (a) Find the change in temperature between sea level and 5 km above sea level.

   (b) Find the temperature at 5 km above sea level.

   (c) At what altitude is it the warmest?

   (d) Sketch the graph of temperature versus altitude.

(e) Indicate the average temperature between 0 and 5 km.

## Function-derivative pairs

In each figure below, a function and its derivative are shown on the same set of coordinate axes. For each figure, identify which graph is the function and which is the derivative, and explain in one or two sentences what properties of the graphs led you to your choice.

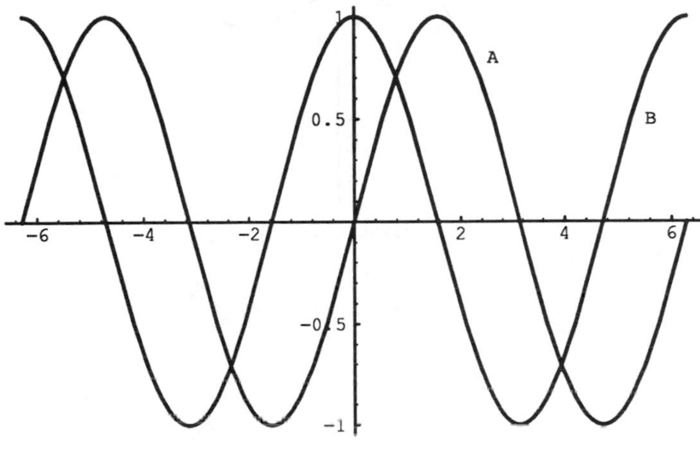


A

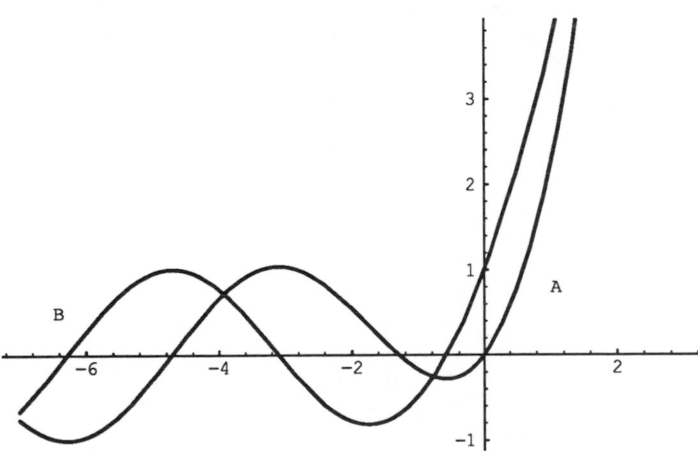


B

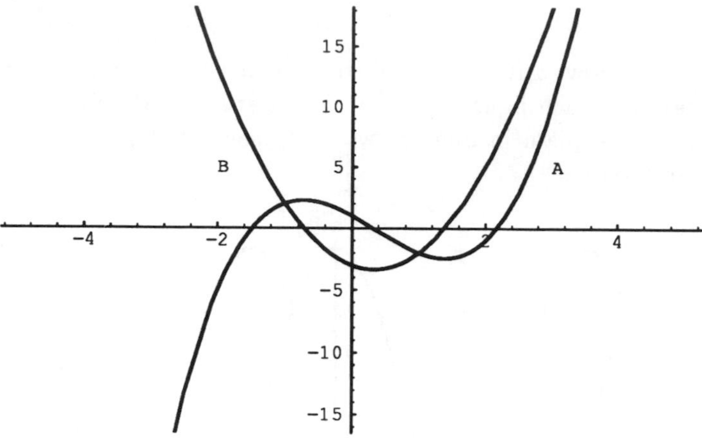


C

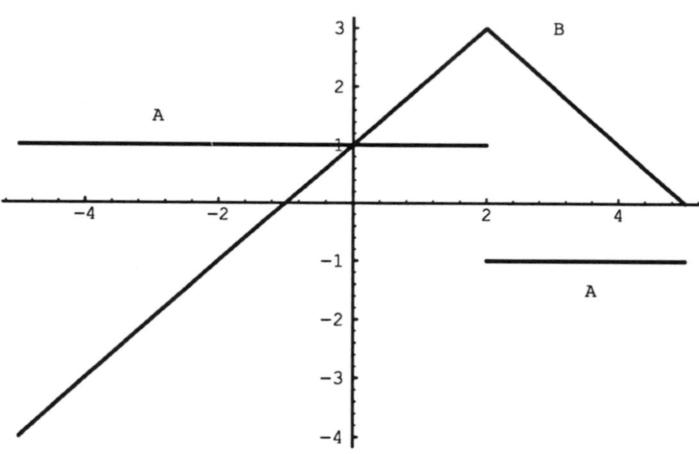


D

## More airplane travel

Let's look again at the plane that passed over Ithaca at 1 o'clock and traveled at 500 mph from 1 to 5. At 5 o'clock it enters the air control region for an airport and starts to slow down. It gradually slows down to 300 mph by 5:30.

1. Draw the velocity graph.

2. Now draw the distance graph.

3. Now the plane lands at 5:45. Draw velocity and distance from 1 to 6.

Two important concepts we will use to analyze graphs are the ideas of *increasing* and *decreasing.* A graph is *increasing* on the interval $(a, b)$ if the graph rises as we move from $a$ to $b$ along the horizontal axis.

4. Write a similar statement describing what it means to say that the graph is decreasing on the interval $(a, b)$.

The distance graph of the plane increased from 1 to 5:45. The velocity graph decreased from 5 to 5:30.

Next let's think of a different story. What if the plane developed engine trouble and turned back towards Ithaca at 2:00?

5. Now what would the distance graph look like?

6. Now try to draw the velocity graph for the last example.

7. Now try to draw the distance graph of the original plane (the plane that did not return to Ithaca) if the plane circled the airport from 5:30 to 5:45.

## Problems

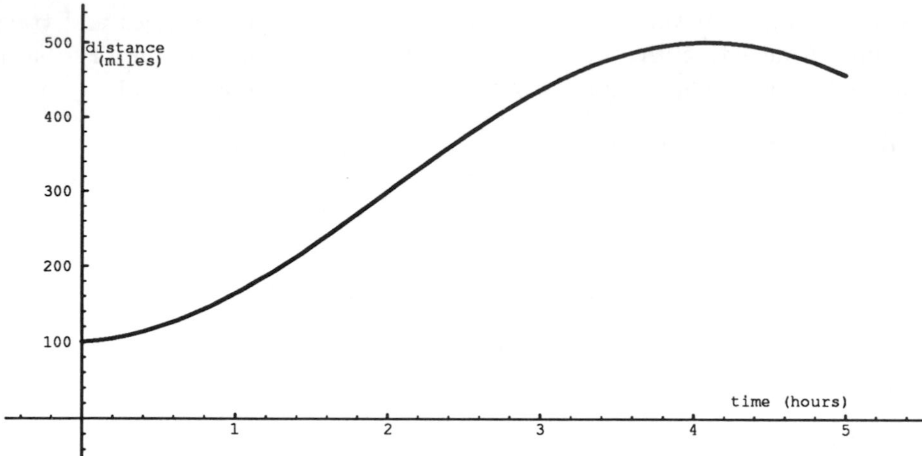


Airplane trip

1. Here is a graph of the distance for another airplane trip. Mark where the velocity is fast and where it is slow. (You can think of the "fast" periods of time as times when the plane enjoys a tailwind, and the "slow" periods as times when there is a headwind.)

2. Dallas to Houston

## Dallas to Houston

The four graphs below each give models for the velocity of a car driving along a straight highway. At time $t = 0$ the car left Dallas and headed for Houston.

1. For each graph, write an explanation of the situation (in complete sentences) that gives the important information contained in the graph.
2. For each graph, decide when velocity is increasing and when velocity is decreasing.
3. Try to give a formula for the velocity in terms of time.
4. Decide whether you think the car could behave like the model. Explain your reasons.
5. Sketch the graph of distance the car has traveled versus time.
6. Try to find a formula for distance in terms of time.
7. For each graph, decide when the car is furthest away from Dallas. Explain.

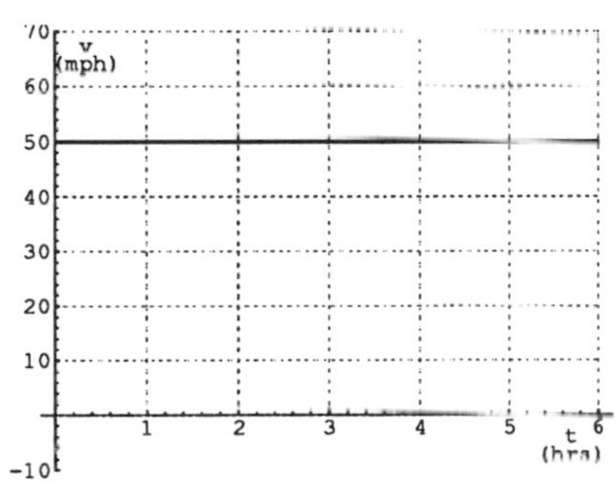


A

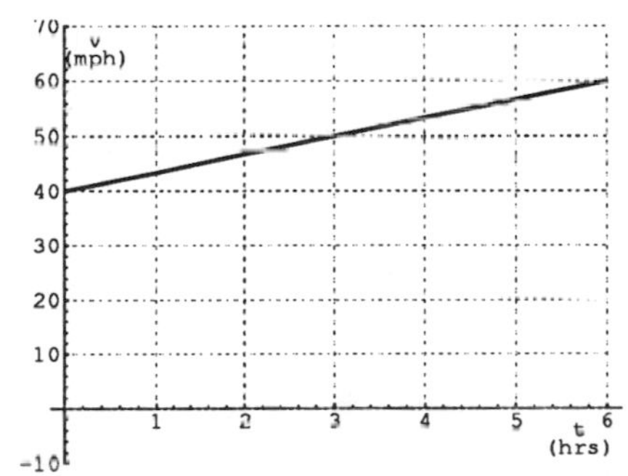


B

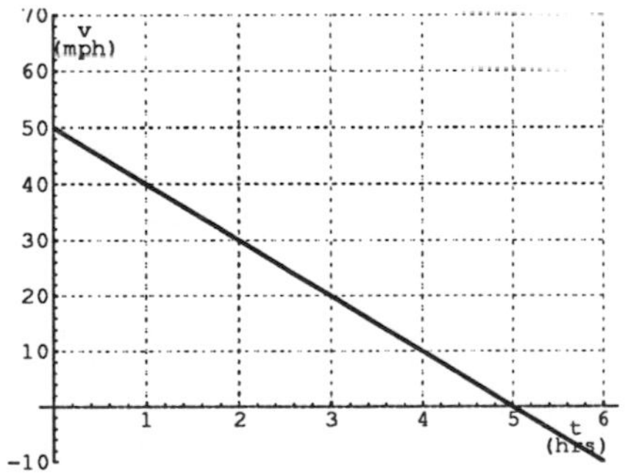

C

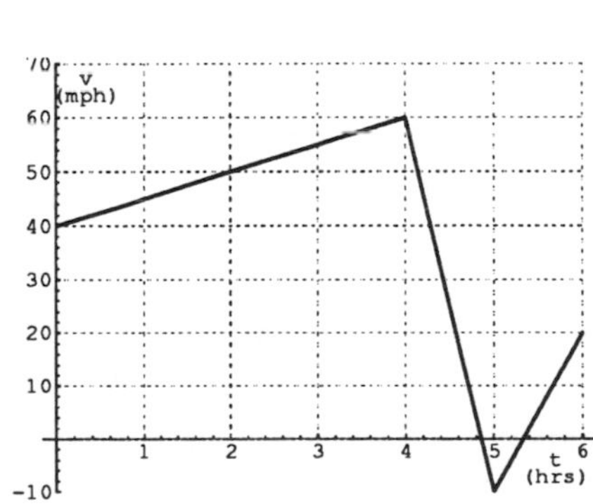


D

## Water tank problem[2]

At time $t = 0$, water begins to flow from a hose into an empty tank at the rate of 40 liters/min. This flow rate is held constant for two minutes, at which time the tank contains 80 liters. At that time, ($t = 2$), the water pressure is gradually reduced, until at time $t = 4$, the flow rate is 5 liters/min. This flow rate is held constant for the final two-minute period, at which time ($t = 6$) the tank contains 120 liters.

1. Draw a graph of the volume $V$ of water in the tank (liters) against the time $t$ (min) for $0 \le t \le 6$.

2. What is the average rate of flow into the tank over the entire six-minute period? Show how this can be interpreted on your graph.

3. Now, suppose that, in addition to the above sequence of events, a pump is started up at time $t = 2$, and, for the next four minutes, pumps water out of the tank at the constant rate of 15 liters/min. What we have called $V$ above now represents the total amount of water that has flowed into the tank at time $t$, and we let $W$ represent the total amount of water pumped out of the tank at time $t$. Plot $W$ on the same set of axes. Show how to interpret the volume of water in the tank at any time.

4. Show how to find on your graph the point at which the water level in the tank is a maximum.

[2] Adapted from Peter Taylor, *Calculus: The Analysis of Functions*, Wall & Emerson, Inc., 1992.

5. What is the instantaneous flow rate from the hose into the tank at this time?

6. Now on a new set of axes draw the graph of the rate at which water enters the tank (liters per minute) against the time (min) for $0 \leq t \leq 6$. (This is the graph of the rate of change of $V$ with respect to time.)

7. On the same set of axes that you used in 6 plot the graph of the rate at which water is pumped from the tank. (This is the graph of the rate of change of $W$.)

8. Find any times where the rate graphs that you drew on parts 6 and 7 intersect. What can you say about the level of water in the tank at these times?

9. If we continue the action past time $t = 6$, with the constant flow in of five liters per minute and the constant flow out of 15 liters per minute, at what time will the tank be empty? Show how to interpret this graphically using the above plot of $V$ and $W$.

## Tax rates and concavity

In this activity, we will use some simple ideas about taxes and tax rates to investigate some basic properties about functions.

Economists use the term *marginal tax rate* to refer to the additional tax paid on the next (taxable) dollar earned. For example, suppose that under some tax plan the tax on \$34,567 is \$6897.76 while the tax on \$34,568 is \$6898.04. Then, for a taxpayer earning \$34,567, the tax paid on the next dollar is \$0.28, so the marginal tax rate is \$0.28/\$1 = 28%. (Note that this is not directly related to \$6897.76/\$34,567 = 19.95%, the percentage of taxable income that goes to tax.)

Let $t(x)$ denote the amount of tax that is paid on $\$x$ of taxable income.

1. Economists refer to a tax as *progressive* if the marginal tax rate increases (as taxable income increases). If $t(x)$ is a progressive tax, what will the graph of $t(x)$ look like? Sketch $t(x)$ for a progressive tax.

2. At a few points on your graph, sketch the tangent line. What is the relationship between the graph and its tangent lines?

3. Explain how the marginal tax rate is related to $t'(x)$. (Hint: Take $\Delta x = \$1$ in the definition of $t'(x)$.)

4. If $t$ has a second derivative, what property will $t''$ have?

A function with an increasing derivative is said to be *concave up.* As you have seen, geometrically this means that the graph lies above its tangent lines; in terms of derivatives, it means that its second derivative is positive.

The definitions above can be "turned upside down"; that is, a function with a decreasing derivative is said to be *concave down.* Geometrically this means that the graph lies below its tangent lines; in terms of derivatives, it means that its second derivative is negative.

**Problems**

Consider the function $u(x) = x - t(x)$.

1. Describe $u$ in words. That is, what does $u(x)$ represent?
2. What properties does $u$ have? (What assumptions are you making about $t$?)
3. Is it possible for a function to be positive, concave down, and unbounded (approach infinity)? Explain.
4. Suppose that over some income range (e.g., for all taxable income between \$21,500 and \$52,000) the marginal tax rate, $t'(x)$, is constant (e.g., 28%). What will the graph of $t(x)$ look like over that range?

## Testing braking performance

Buyers' Union is testing the braking performance of two cars: the Minima and the Corollari. The test consists of using the brakes to bring the car to a complete stop when the car is traveling at 60 miles per hour. The Minima slows down at the uniform rate of 8 miles per hour per second. The Corollari's velocity is given in the graph below.

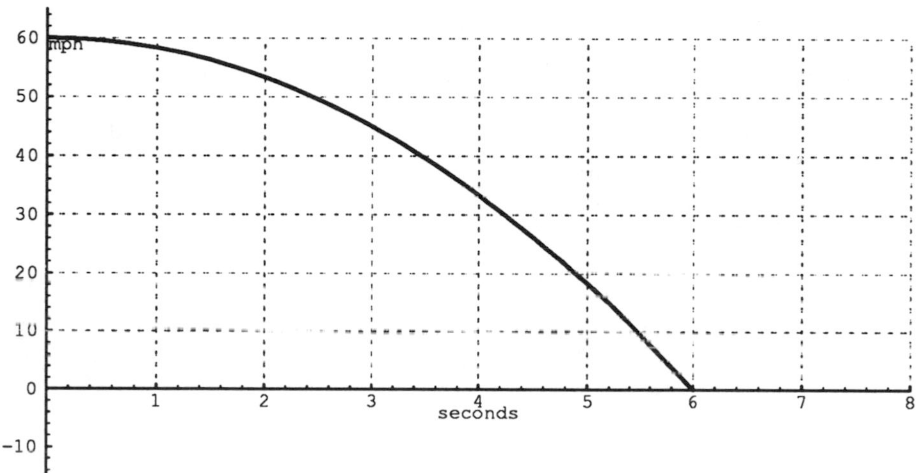


Velocity curve

1. How much time will it take for each car to stop?

2. How far will each car travel before it stops?

3. Write a brief paragraph summarizing braking performance which will be inserted into the report comparing these cars. If your data indicates that one of the cars is superior explain your conclusion.

## The start-up firm

Pat and Chris are involved in a new business. One of Pat's jobs is to estimate the company's prospects for raising money (finding investors at the beginning, selling their product eventually); and Chris is responsible for spending it (buying equipment, finding office space, hiring staff, ...). Both have been making predictions about how much they will be raising/spending. What is shown on the graph below is a graph of these projections. The horizontal axis represents months; the vertical axis measures dollars per month. (Note that what is illustrated is the *rate* at which money is raised/spent—not the total amount raised/spent.) Both are particularly interested in the company's net worth (the difference between the amount raised and the amount spent).

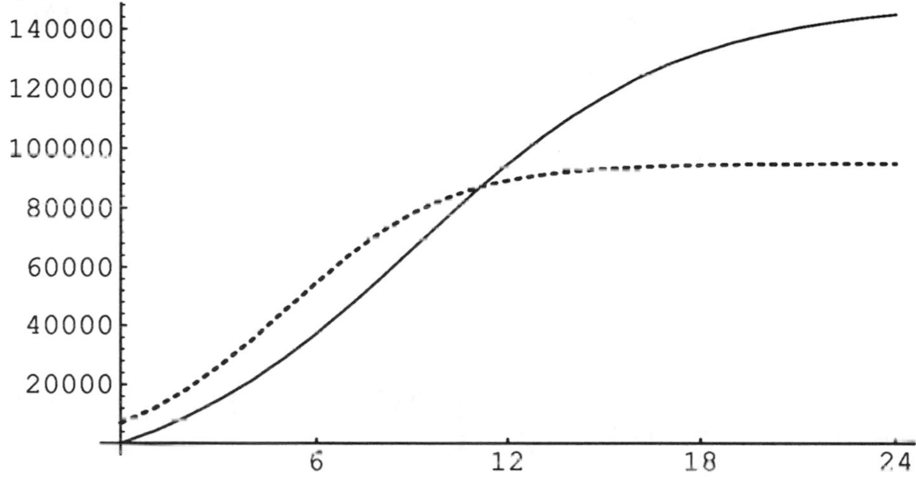


Pat: solid line; Chris: dotted line

1. Estimate the company's net worth in two months.

2. Is net worth positive or negative after six months? after 12 months? after 24 months?

3. When is net worth decreasing? When is it increasing?

4. When is net worth smallest? largest? zero?

5. Sketch a graph of net worth.

6. Both graphs seem to flatten out. Interpret this phenomenon. (What does this tell you about net worth?)

## Graphical composition

Below is a sketch of the graphs of two functions $y = f(x)$ and $y = g(x)$. Given a point $x_0$ in the domain of $f$ with the property that $g(x_0)$ is in the domain of $f$, there is a geometric method that can be used to plot $(x_0, f \circ g(x_0))$. The method determines a polygonal path (a path made up of straight line segments) that starts at the point $(x_0, 0)$ on the $x$-axis and ends at $(x_0, f \circ g(x_0))$. We illustrate the construction of this path below.

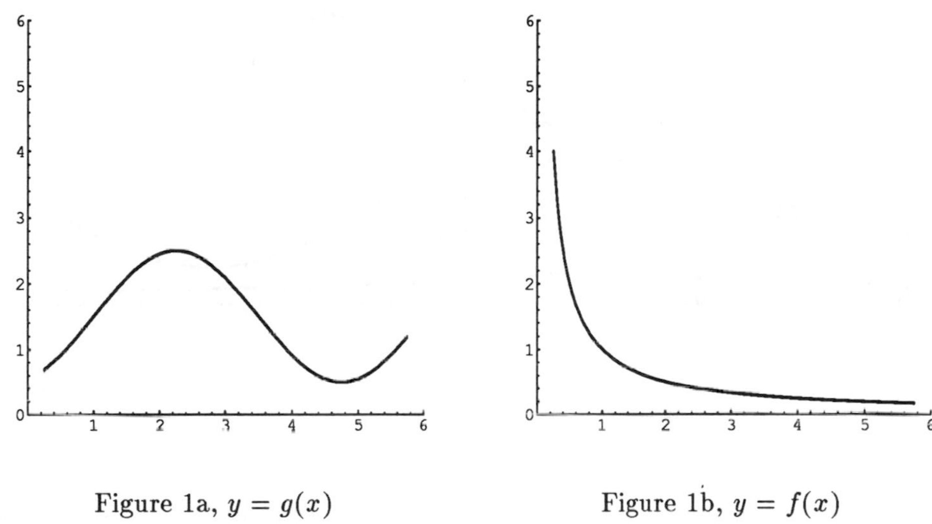

Figure 1a, $y = g(x)$ Figure 1b, $y = f(x)$

First, superimpose the graph of $f$ on the graph of $g$ and also sketch the graph of the line $y = x$.

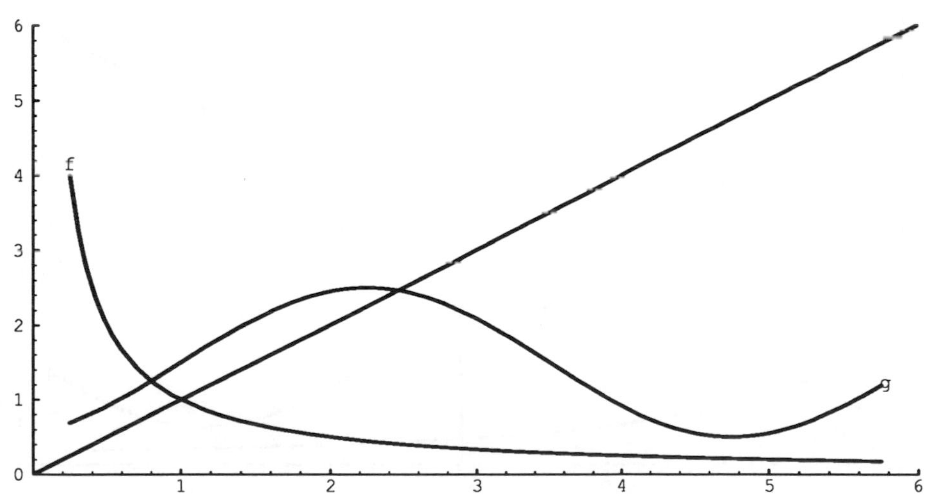


Figure 2

Next, starting at the point $(x_0, 0)$, travel along the line $x = x_0$ until you meet the graph of $g$ at $(x_0, g(x_0))$. This determines the first leg of the path. (See Figure 3.)

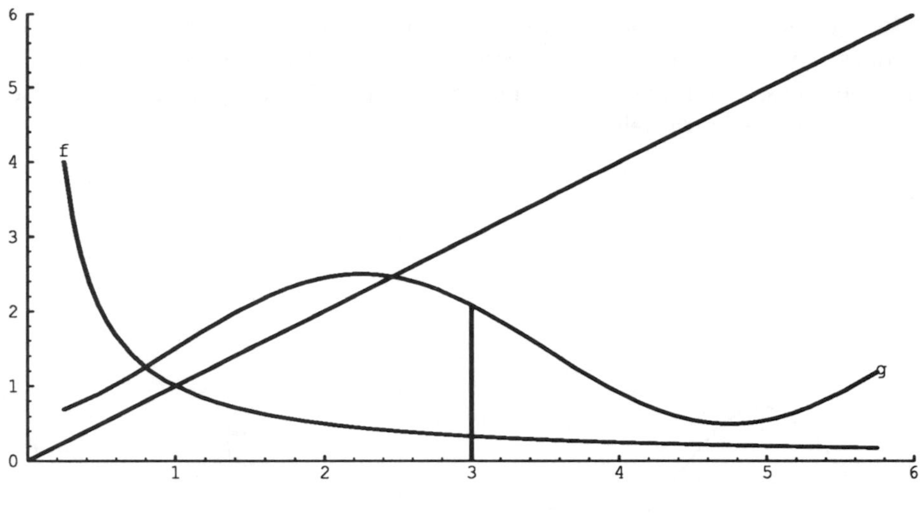


Figure 3

Now, travel along the line $y = g(x_0)$ until you meet the line $y = x$ at the point where the $x$ and $y$ coordinates are equal. This is the point $(g(x_0), g(x_0))$. This determines the next leg of the path. (See Figure 4.)

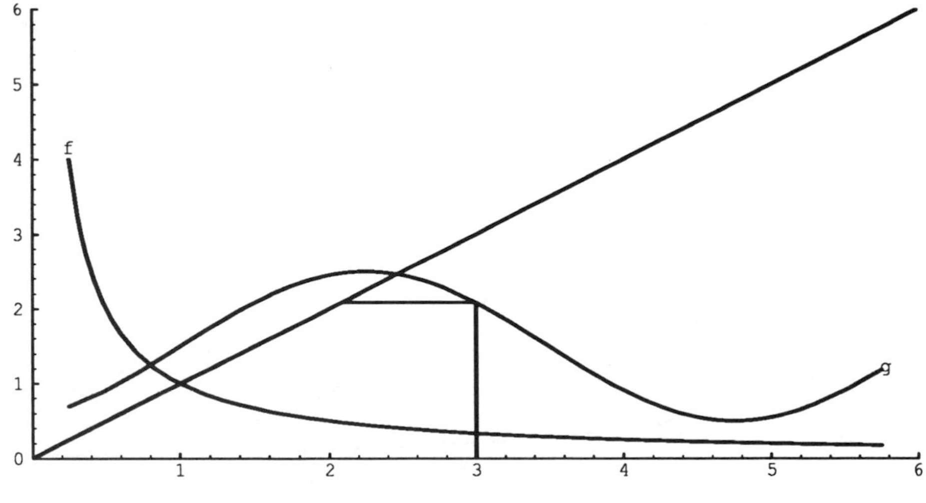


Figure 4

Next, travel along the line $x = g(x_0)$ until you meet the graph of $f$ at $(g(x_0), f(g(x_0)))$. This determines the third leg of the path. (See Figure 5).

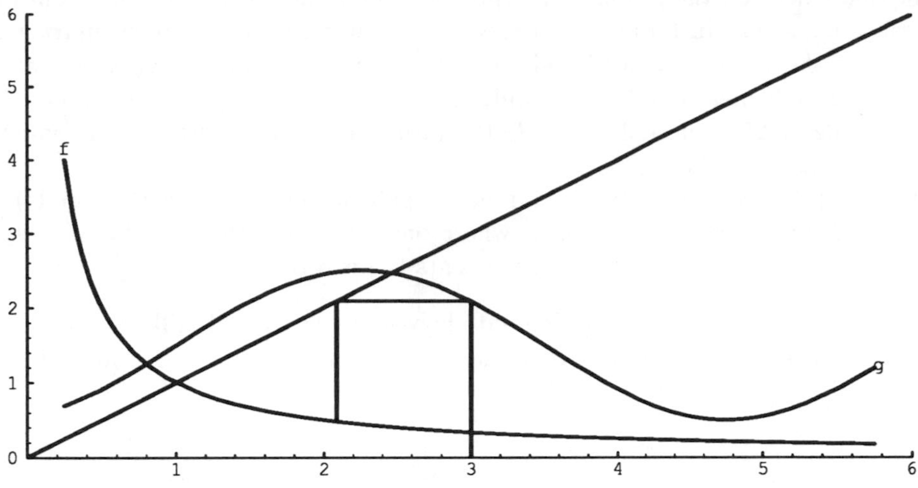


Figure 5

Finally, travel along the line $y = f(g(x_0))$ until you meet the line $x = x_0$ at $(x_0, f(g(x_0))$. This is the last leg of the path and it terminates at $(x_0, f(g(x_0))$, the desired point. (See Figure 6.)

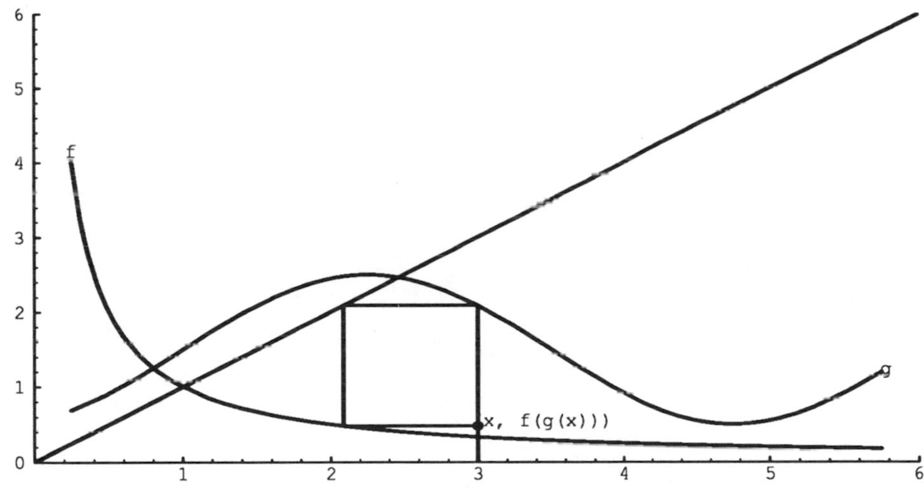


Figure 6

To recap, start at $(x_0, 0)$, travel along $x = x_0$ to the graph of $g$, travel to the line $y = x$, travel to the graph of $f$, travel back to the line $x = x_0$.

For the functions graphed above:

1. Find $(f \circ g)(2)$.
2. Find $(f \circ g)(4)$.

**Problems**

In this example, the graph of $f$ represents the number of ten-gallon containers of fresh milk a certain farmer can expect to sell per day on the open market as a function of the total amount of milk available in a particular metropolitan market. The total milk is measured in thousands of gallons. For example, if only 250 gallons are available (0.25 thousand gallons), the farmer could expect to sell 4 ten-gallon containers; i.e., 40 gallons.

The graph of $g$ represents the number of gallons (measured in thousands) as a function of time over the six months where one represents the end of January, two represents the end of February, six represents the end of June, etc.

1. (To be done individually.) Estimate how many gallons of milk the farmer was able to sell on the last day of January, of February, of March, and of June.

2. (Group) Can you give a sketch of the farmer's selling potential over the time scale $[1, 6]$? Give a description of what this graph says about the farmer's selling potential.

# The leaky balloon

A spherical balloon is leaking. Below is a graph of the radius ($r$) as a function of time ($t$).

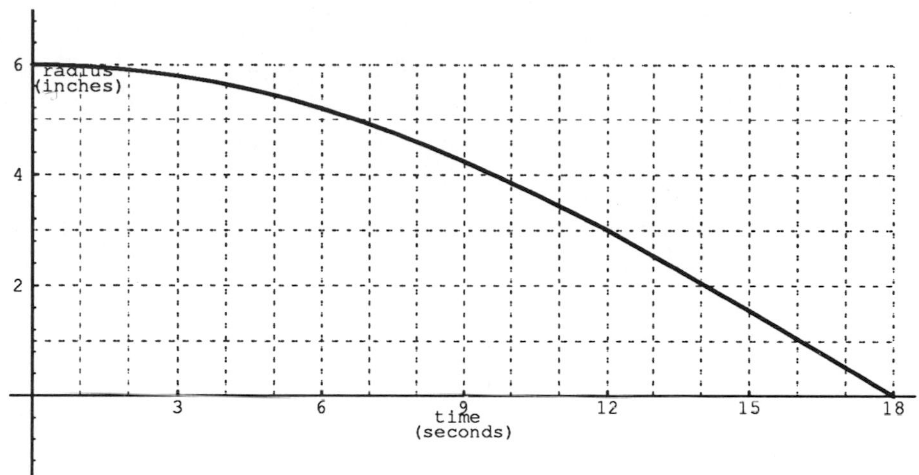


Leaky balloon

1. How fast is the radius changing when $t = 15$?

2. How fast is the radius changing when the radius is 4 inches?

3. How fast is air escaping (i.e., how fast in the volume changing) when the radius is 4 inches? [$V = \frac{4\pi r^3}{3}$.]

**Problem**

How fast is air escaping when $t = 15$?

## Inverse function from graphs

Figure 1 refers to a graph of the velocity of an airplane during the first phase of its trip. Here, the plane is traveling at different velocities at different instants. So, if we had a record of the velocities, we could theoretically pinpoint the one and only one instant that the plane was traveling at that velocity. The function that determines the time from the velocity is called the *inverse* of the function $v$ and is denoted $v^{-1}$. So, in the terminology of inverse functions, for each time $t_0$, $v^{-1}(v(t_0)) = t_0$, and for each velocity $v_0$, $v(v^{-1}(v_0)) = v_0$. Obviously $v$ would not have an inverse if the plane took on the same velocity at more than one instant, for in that case we couldn't determine a unique time from that velocity.

Using the fact that $v^{-1}(v(t_0)) = t_0$ and our geometric results on composition, we can sketch the graph of $v^{-1}$ on the *same* axis as the graph of $v$.

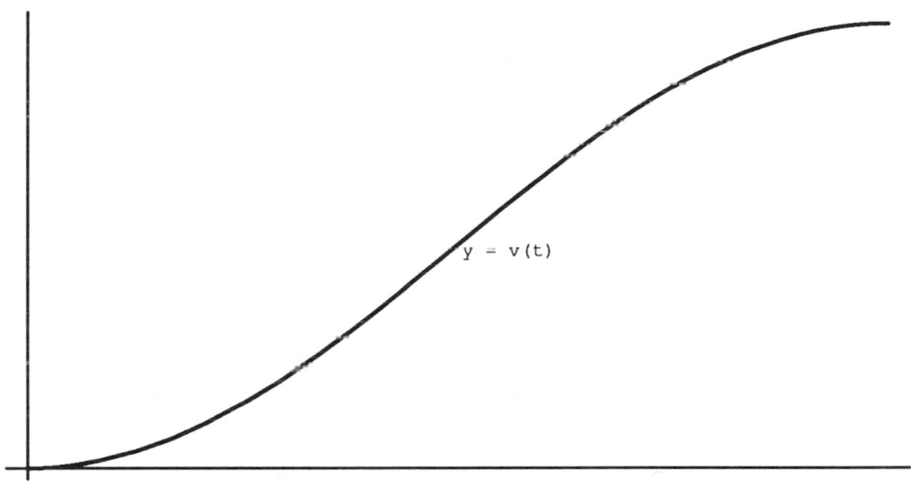


Figure 1

Note that for the function $v$, the $x$-axis is interpreted as a time axis, while when we consider the function $v^{-1}$, the points on the $x$-axis measure velocity.

1. Give an argument that supports the claim that given $v_0$, the end of the polygonal path in Figure 2 will be the point $(v_0, v^{-1}(v_0))$.

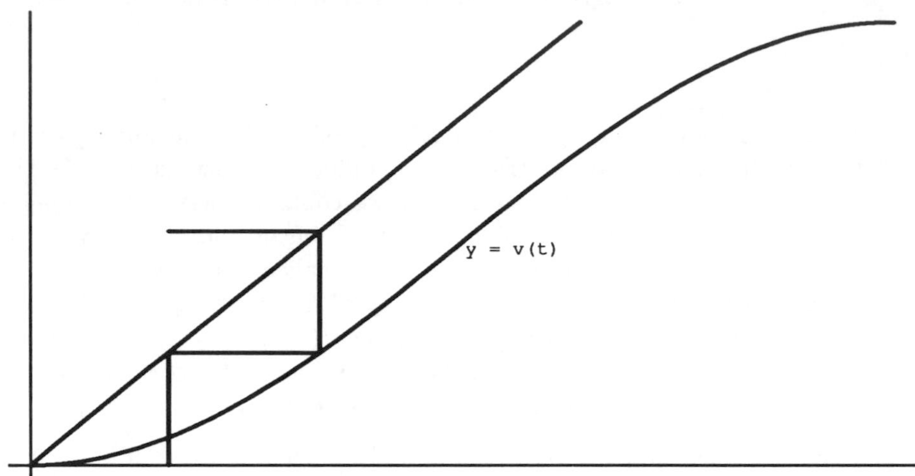


Figure 2

2. Give an argument that supports the claim that the box in Figure 3 is a square.

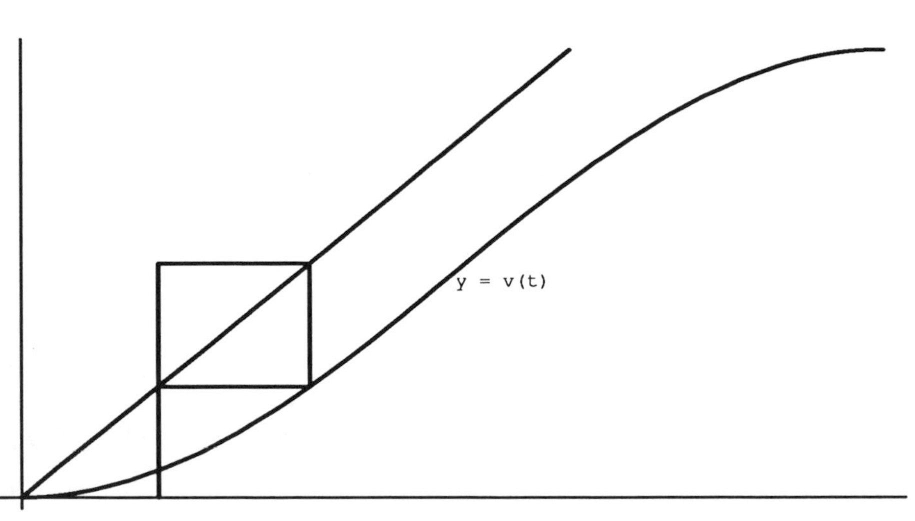


Figure 3

3. Give an argument that suggests that $(v_0, t_0)$ is the reflection of the point $(t_0, v_0)$ about the line $y = x$.

4. Using the results of 3, sketch a graph of $v^{-1}$ on the same axes in Figure 1. (Plot a few points of $v^{-1}$ first.)

## Problems

1. Sketch the graph of an increasing function $y = f(x)$.
2. Sketch a graph of $f^{-1}$.
3. Sketch a graph of $f^{-1} \circ f$.

# Chapter 2

# Functions, Limits, and Continuity

## Introduction to functions

One of the most useful ideas in mathematics is the concept of function. So far in these pages many examples of functions have occurred, though we have not especially stressed the properties that distinguish the examples as functions.

A *function* is a rule that associates with each element of some set, called the *domain* of the function, exactly one element of another set, called the *co-domain* of the function. If the name of the function is $f$, if its domain is $X$, and if its co-domain is $Y$, then we convey these facts by using the notation

$$f: X \to Y.$$

In first-year calculus most functions have as both domain and co- domain the set $\mathbf{R}$ of all real numbers, or some subsets of $\mathbf{R}$.

The way in which the function associates elements of the domain with elements of the co-domain can be conveyed in a variety of ways: by a graph, in words, by a table or a formula; functions often arise from physical phenomena. The important thing to remember is that in order to be a function, for *each element* of the domain $X$, there must be *exactly one element* of the co-domain $Y$.

For example, consider the graph of Josh's interrupted trip to the library, in Library Trip. It represents a function, which we will call $G$. The graph is reproduced here.

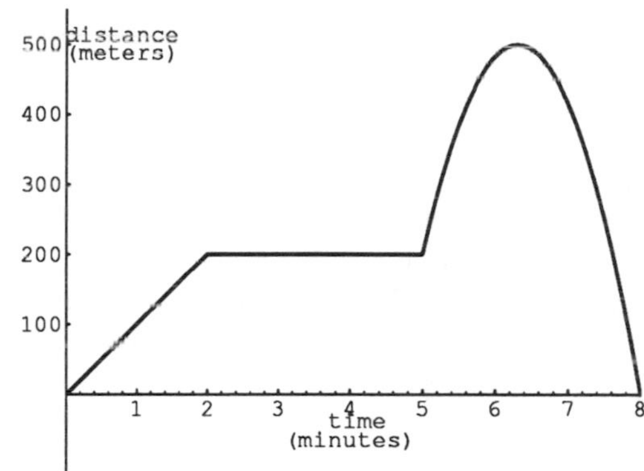


Library trip

When a function is presented on a graph, it is customary to represent the domain of the function on the horizontal axis and to represent the co-domain on the vertical axis. The domain is the part of the horizontal axis that corresponds to points actually included in the graph. In this example, the domain of $G$ consists of all the numbers between 0 and 8. In set notation, we would write

$$\text{domain of } G = \{x \in \mathbf{R} \mid 0 \le x \le 8\}$$

We can think of the co-domain as the set $\mathbf{R}$ of all real numbers. However, not every real number on the vertical axis corresponds to a point on the graph. The elements of the co- domain that actually correspond to points on the graph make up the *range* of the function. In this case, the range of the function $G$ is the set

$$\text{range of } G = \{y \in \mathbf{R} \mid 0 \leq x \leq 500\}$$

To determine what element of the range is associated with a specific element of the domain, we would read the value from the graph. Since the points $(2, 200)$ and $(6, 483)$ lie on the graph, we see that the function $G$ associates with the number 2 (in the domain) the number 200 (in the range), and with the number 6 (in the domain) $G$ associates the number 483 (in the range). It is customary to use more concise notation to express these facts. For the two facts above, we would write

$$G(2) = 200 \quad \text{and} \quad G(6) = 483$$

1. Use the graph to find (or estimate) $G(4)$, $G(0.7)$, $G(5.9)$, and $G(0)$.

2. Given that a function must associate exactly one range element with each domain element, how can you tell by looking at a graph whether or not the graph represents a function?

3. Which of the graphs below represent functions? Explain.

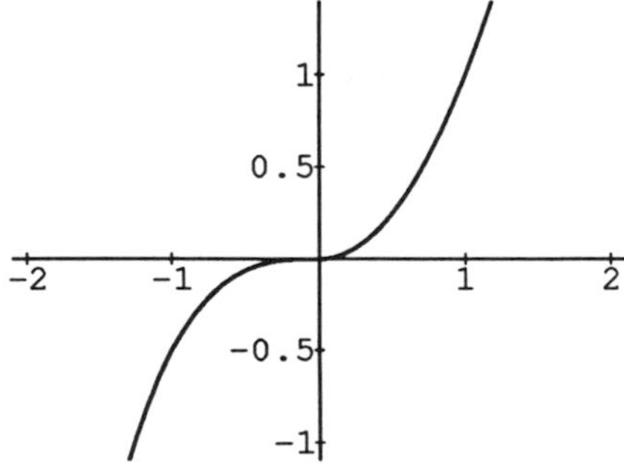


A

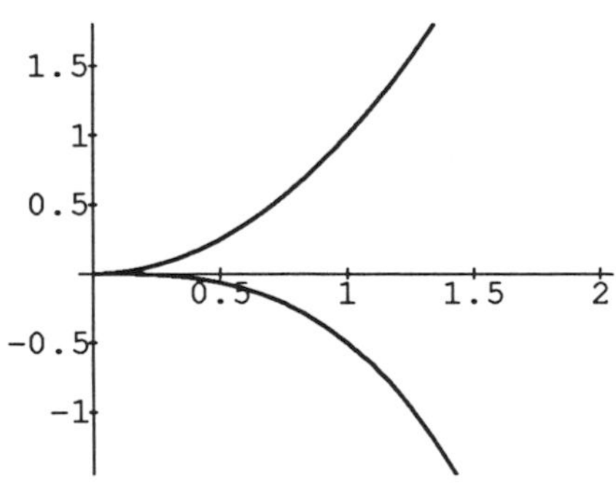


B

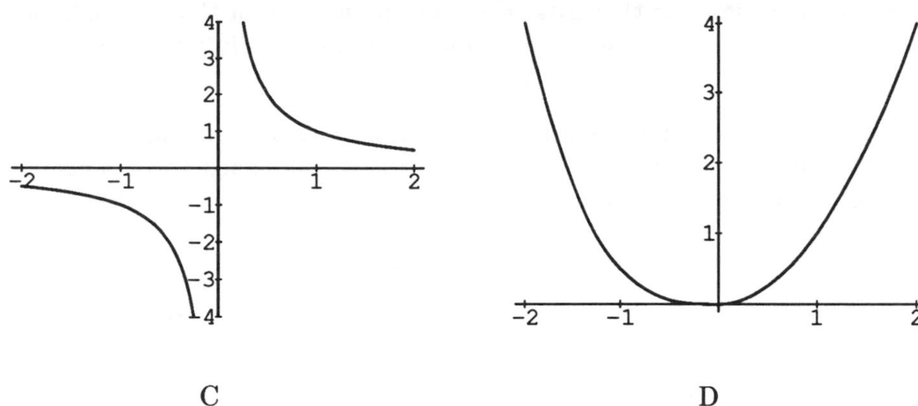

C D

4. If $T$ is a function that is presented as a table, we read the function values directly from the table. For example, the table at the right represents a function. The domain is the set of integers $\{1, 2, 3, 5\}$.

| $x$ | $T(x)$ |
|---|---|
| 1 | 13 |
| 2 | 4 |
| 3 | -6 |
| 5 | 0.1 |

For this function, you can read the values right off the table. For example, $T(3) = -6$. Find $T(1)$, $T(2)$, and $T(5)$.

Notice that it doesn't make sense to ask for $T(4)$ or $T(8.7)$ because 4 and 8.7 are not in the domain of the function. Also notice that if we added a line "2 7" to the table we would have violated the "one and only one value" rule by trying to have two different values associated with 2.

5. Some functions are presented as formulas. If this suits the situation, this can be very useful, because there are lots of well understood methods of analyzing the properties of such functions. For example, the formula $f(x) = 3x + 4$ defines a function. Unless explicitly stipulated, the domain consists of all the values of $x$ that make sense in the formula. The domain in this case is all real numbers. For any domain element, the formula tells us the way to calculate a range element. Thus, $f(-0.8) = 1.6$ (i.e., $3 \times -0.8 + 4$). Find $f(2)$, $f(0.06)$, $f(-\pi)$.

For such functions, we usually write $f(x) =$ formula in $x$. This notation tells you the name of the function (in this case, $f$) and helps clarify what the domain is. (As mentioned above, unless otherwise specified, the domain consists of all the values of $x$ that make sense in the formula on the right side of the equals sign.) We obtain specific function values, as we did above, by substituting for x on both the left and the right side.

One way of visualizing a function is as a machine. The machine accepts inputs (values from the domain), does whatever it was designed to do, and then delivers outputs (members of the co-domain).

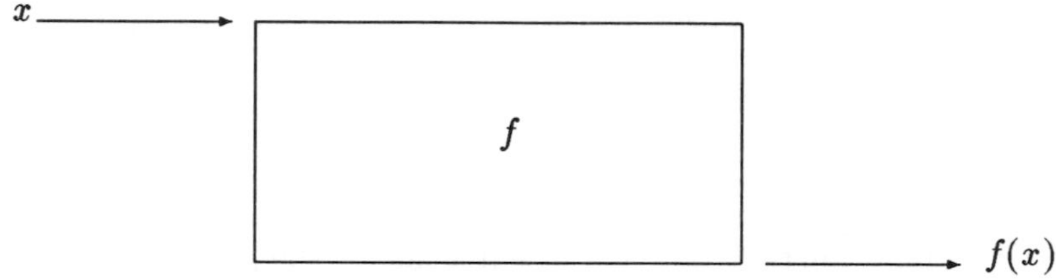


Inside the machine can be a formula, a table, a graph, or some other mechanism. As long as the machine produces exactly one output for every (valid) input, it represents a function.

**Problem**

Find the domain and range for all the functions you found in 3.

## Postage

You are in charge of a mail order company. Your company sends out 10,000 letters each day. The post office rates for postage are:

29 cents for a letter that weighs 1 ounce or less,

54 cents for a letter that weighs more than 1 ounce but not more than 2 ounces,

77 cents for a letter that weighs more than 2 ounces but not more than 3 ounces, and

98 cents for a letter that weighs more than 3 ounces but not more than 4 ounces.

1. How much will it cost to mail a letter that weighs 1 ounce? 1.5 ounces? 0.8 ounces?

2. Sketch a graph of the postage for a letter as a function of the weight of the letter for weights from 0 to 4 ounces.

3. The catalog department plans to send out 10,000 letters that each weigh .85 ounces. How much will the postage cost?

4. The marketing department wants to add an insert to these letters. The insert weighs 0.2 ounces. If you agree to add the inserts to the letters, what will happen to the cost of postage for the letters?

5. The book department is sending out 5,000 letters that each weigh 0.75 ounces. What will happen to the price of the book department's mailing if you agree to add the marketing department's insert to each letter?

6. Summarize what the graph in 2 means to the head of the mail order company.

## Problems

1. A long distance telephone call costs $1.25 for the first minute or less and $0.50 for every additional 30 seconds or part of 30 seconds after the first minute.

   (a) How much will it cost to talk for 2 minutes? for 2 minutes and 5 seconds? for 2 minutes and 25 seconds?

   (b) Draw the graph of the cost of a long distance call as a function of the length of the call for calls that last from one second to four minutes.

   (c) Point out any discontinuities on your graph and explain what they mean to a student who calls long distance every night.

2. A parking lot charges $2 for the first hour or less and then $1.50 for any hour or fraction of an hour after the first hour.

   (a) Sketch the graph of the parking charge as a function of how long a car remains in the lot for any time up to 5 hours.

   (b) Point out any discontinuities on your graph and explain what they mean to someone who parks in the lot every day.

## What's continuity

If you look at the graph of the function below, you can probably say exactly where the function $f$ is and is not continuous.

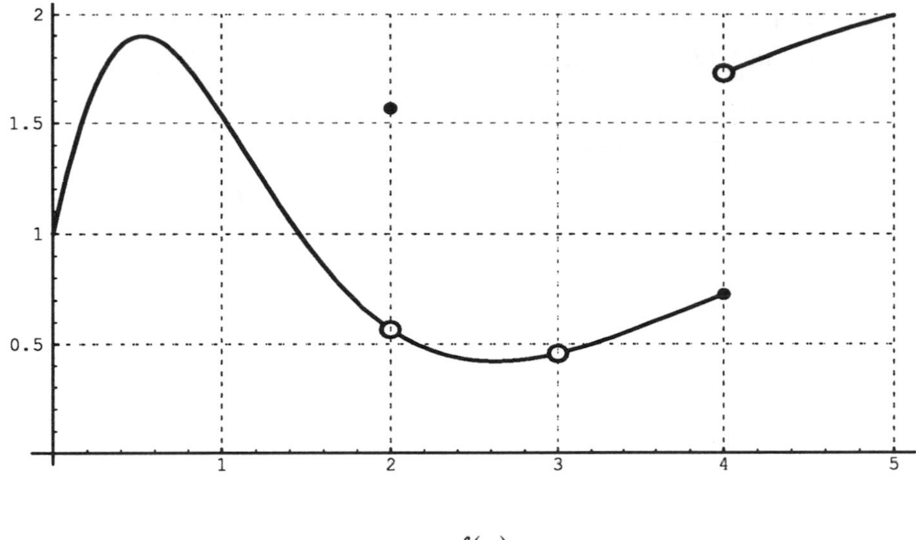


$y = f(x)$

1. For what values of $x$ is $f$ *not* continuous?

2. For each of the following values of $c$, indicate $\lim_{x \to c} f(x)$, $f(x)$, and whether $f$ is continuous at $x = c$.

   (a) $c = 1$

   $\lim_{x \to 1} f(x) =$ ____________

   $f(1) =$ ____________

   Is $f$ continuous at $x = 1$? ______

   (b) $c = 2$

   $\lim_{x \to 2} f(x) =$ ____________

   $f(2) =$ ____________

   Is $f$ continuous at $x = 2$? ______

(c) $c = 3$

$\lim_{x \to 3} f(x) =$ ____________

$f(3) =$ ____________

Is $f$ continuous at $x = 3$? ______

(d) $c = 4$

$\lim_{x \to 4} f(x) =$ ____________

$f(4) =$ ____________

Is $f$ continuous at $x = 4$? ______

3. Using the language of limits, what does it mean for a function $f$ to be continuous at $c$?

## Limits and continuity from a graph

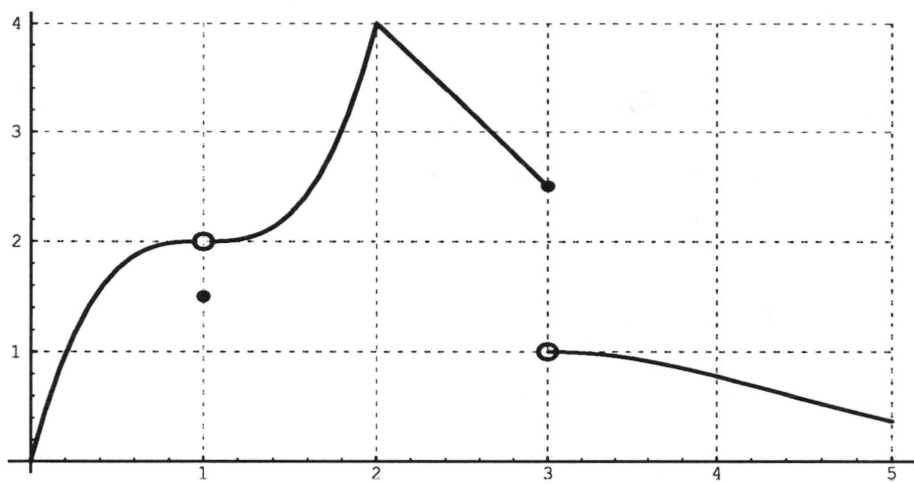


Graph of the function $y = f(x)$

1. Find the value of each of the following. If the value does not exist, (undefined), write **DNE**.

   (a) $\lim_{x \to 1^+} f(x)$

   (b) $\lim_{x \to 1^-} f(x)$

   (c) $\lim_{x \to 1} f(x)$

   (d) $f(1)$

   (e) $\lim_{x \to 2^+} f(x)$

   (f) $\lim_{x \to 2^-} f(x)$

   (g) $\lim_{x \to 2} f(x)$

   (h) $f(2)$

   (i) $\lim_{x \to 3^+} f(x)$

   (j) $\lim_{x \to 3^-} f(x)$

   (k) $\lim_{x \to 3} f(x)$

   (l) $f(3)$

   (m) $\lim_{x \to 4^+} f(x)$

(n) $\lim_{x \to 4^-} f(x)$

(o) $\lim_{x \to 4} f(x)$

(p) $f(4)$

2. True or false, with reasons:

   (a) The function $f$ is continuous at $x = 1$.

   (b) The function $f$ has a derivative at $x = 1$.

   (c) The function $f$ is continuous at $x = 2$.

   (d) The function $f$ has a derivative at $x = 2$.

   (e) The function $f$ is continuous at $x = 3$.

   (f) The function $f$ has a derivative at $x = 3$.

   (g) The function $f$ is continuous at $x = 4$.

   (h) The function $f$ has a derivative at $x = 4$.

## Slopes and difference quotients

Answer the following questions about the function $f$, whose graph is given below.

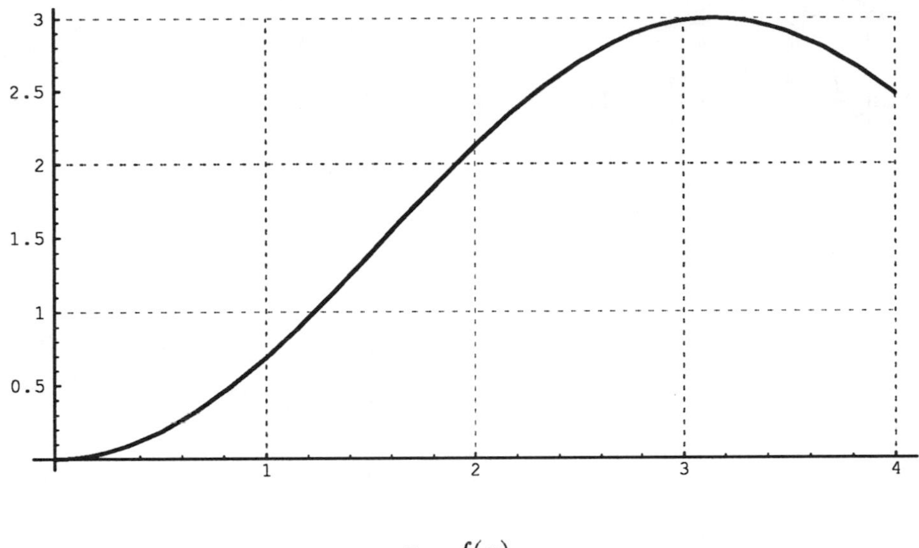


$y = f(x)$

1. Find $f(3)$.

2. Find $f'(3)$.

3. What is the average rate of change of $f$ over the interval $[0, 3]$.

Sketch on the graph a line whose slope equals:

4. $\dfrac{f(3) - f(2.5)}{3 - 2.5}$

5. $\dfrac{f(3)}{3}$

6. $\lim\limits_{h \to 0} \dfrac{f(3 + h) - f(3)}{h}$

7. $\lim_{h \to 0} \frac{f(h)}{h}$

Mark a point on the curve where the $x$–coordinate has the property:

8. $\frac{f(x)}{x} = 1$

9. $\lim_{h \to 0} \frac{f(x+h) - f(x)}{h} = 1$

## Sequences

1. Using the $\boxed{x^2}$ button on your calculators, the group should construct terms of several sequences using the following procedure. Each member of the group should chose a different number and record it. This number should be considered the first term of a sequence and entered into the calculator. The $\boxed{x^2}$ button should be pushed several times and each result recorded as the next term of the sequence until a pattern of convergence or divergence develops. If the sequence appears to be approaching a limit, specify the limit. If the sequence appears to be diverging indicate this also. Each member of the group should repeat this process with two *additional* initial numbers. Now, as a group, consider the following questions. When will you get a convergent sequence? How many different limits are there? What happens to the difference between successive terms when the sequence converges? diverges?

2. Repeat the above procedure using the $\boxed{\sqrt{x}}$ button on the calculator.

3. Discuss the similarities and differences in results obtained in 1 and 2 above.

4. You have constructed several terms of several sequences associated with functions $f(x) = x^2$ and $g(x) = \sqrt{x}$. For the sequences that diverged, can you find a relationship between the derivative and the difference between successive terms of the sequence? What about a relationship between the derivative of the function at points of a convergent sequence and the difference between successive terms of the sequence? Is there any relationship between the derivative of a limit point of a sequence and the kind of sequences that converge to it?

## Can we fool Newton?

In class you have discussed Newton's method for finding a zero of a given differentiable function $f$. The method consists of computing terms of a sequence $\{x_n\}$ by choosing an initial value $x_0$ and determining the values $x_1$, $x_2$, $x_3$, and so on using Newton's "formula." If the sequence $\{x_n\}$ converges to a point $x_*$ in the domain of the function, the point $x_*$ might be a root of the equation $y = f(x)$. This activity explores some of the questions concerning the sequence $\{x_n\}$ . Most questions deal with cases when things go wrong.

1. If one exists, sketch a portion of the graph of a differentiable function $f$ such that there is a point $x_0$ where the sequence $\{x_n\}$ determined using Newton's method diverges to infinity. Give a broad description of your function. If no such function exists, support this conclusion.

2. If one exists, sketch a portion of the graph of a function $f$ for which there is exactly *one* point $x_0$ in the domain of $f$ such that the corresponding sequence $\{x_n\}$ determined by Newton's method converges. If no such function exists support your conclusion.

3. Suppose $\{x_n\}$ has been determined by Newton's method.

(a) If $x_3 = x_0$, discuss whether or not $x_4 = x_1$.

(b) Sketch a portion of the graph of a function for each of the following:

i. There is a point $x_0$ with $x_0 = x_2$, but $x_0 \neq x_1$.

ii. There is a point $x_0$ with $x_0 = x_3$, but $x_0 \neq x_2$.

(c) Describe a general process for determining a function $f$ that has the property that there is a point $x_0 = x_n$, but $x_0 \neq x_{n-1}$, for $n = 4, 5, 6, \ldots$

**Problem**

Having had the experiences 1–3, discuss the types of behavior that might happen to a sequence $\{x_n\}$ determined by Newton's method for a differentiable function $f$. What behavior would you expect if $x_0$ is randomly chosen and $f$ is a trigonometric function? if $f$ is an exponential function?

# Chapter 3

# Derivatives

## Linear approximation

In "Graphical estimation of slopes," you learned to estimate the slope of a graph by looking at it "close up," and observing that it resembles a straight line locally. Actually, the graphs you examined in that worksheet never really get perfectly straight. They just look that way from close-up. However, we can say that there are straight lines that approximate this part of the graph when you look at very small pieces. And among these approximating lines, there is one that is the best linear approximation to the graph near the point on which you are focusing your attention. Intuitively this means that no matter how close you get to the curve at this point, this "best" line still looks like a good approximation.

In this exercise, you will look at a close up view of a graph together with its best linear approximation at a point. Then you will "back off" and view the relationship of the entire line to the entire original graph.

For the first view, graph the functions $f(x) = x/(1+x^2)$ and $L(x) = 0.48x-0.16$, using a viewing window from about $x = -0.505$ to $x = -0.495$ and $y = -0.405$ to $y = -0.395$. The graph you see is a close up view of $f(x)$ near the point $(-0.5, -0.4)$, together with its best linear approximation, $L(x)$. Notice that you really see only one figure there. The graph of $f$ and the graph of its best linear approximation appear to coincide (lie on top of each other) from this close in.

Now change the domain and range of the graph three times as follows (the given values are the new domain and range):

1. From $x = -0.55$ to $x = -0.45$ and from $y = -0.45$ to $y = -0.35$
2. From $x = -1$ to $x = 0$ and from $y = -0.9$ to $y = 0.1$
3. From $x = -3.5$ to $x = 2.5$ and from $y = -3.4$ to $y = 2.4$

From what you have observed, you should have a pretty good idea what the best linear approximation to a graph looks like. The best linear approximation to a graph at a given point is called the *tangent line to the graph at the point.* The slope of the tangent line, which we identified in the activity "Graphical Estimation of Slope" as the slope of the curve at the given point, is called the *derivative of the function at the given point.*

### Problems

1. "Slope with rulers" activity

## Estimating cost

A clothing company makes dress shirts. The table below gives information about the total cost of making shirts on a typical day. The top line of the table gives the total number of shirts made since the beginning of the day and the bottom line gives the cost (in dollars) of making those shirts.

| Number | 0 | 10 | 20 | 30 | 40 | 50 | 60 | 70 | 80 |
|---|---|---|---|---|---|---|---|---|---|
| Cost | 1000 | 1316 | 1447 | 1548 | 1632 | 1707 | 1775 | 1837 | 1894 |

1. (a) What is the average cost of a shirt for the last 40 shirts that were made?

   (b) What is the average cost of a shirt for the last 20 shirts that were made?

   (c) What is the average cost of a shirt for the last 10 shirts that were made?

2. Based on the information you gathered in 1, do you think the graph of the cost near $(80, 1894)$ is increasing and concave up, increasing and concave down, decreasing and concave up, or decreasing and concave down? Explain your choice.

3. The company is trying to estimate how much it will cost to make one more shirt (in addition to the 80 already made) .

   (a) Can the company make another shirt for less than \$10? Explain.

   (b) Can the company make another shirt for less than \$6? Explain.

   (c) What is your best estimate for the cost of one more shirt? Explain.

**Problems**

Quarter horses race a distance of 440 yards (a quarter mile) in a straight line. During a race the following observations were made. The top line gives the time in seconds since the race began and the bottom line gives the distance (in yards) the horse has traveled from the starting line.

| Time | 2 | 4 | 6 | 8 | 10 | 12 | 14 | 16 | 18 | 20 |
|---|---|---|---|---|---|---|---|---|---|---|
| Distance | 30 | 64 | 100 | 138 | 177 | 217 | 259 | 302 | 347 | 397 |

1. What is your best estimate for how fast the horse is running halfway through the race?

2. The horse will win a bonus if the time for the race is less than 22 seconds. Decide whether you think the horse will win the bonus. Explain your reasons.

# Finite differences

## Part A

1. Each person in the group should choose a different polynomial (preferably of differing degrees $\leq 3$) and complete the table below as follows. If your polynomial is $y = p(x)$, in Row 1 compute $p(1)$, $p(2)$, $p(3)$, $p(4)$, $p(5)$, $p(6)$. In Row 2, over the star "*" compute $(p(2) - p(1))$, $(p(3) - p(2))$, $(p(4) - p(3))$, $(p(5) - p(4))$, and $(p(6) - p(5))$. In Row 3 and in subsequent rows, over the stars, compute the difference between the entry in the row directly above the star to the right and the entry in the row directly above the star to the left. See Example 1.

| 1 | | 2 | | 3 | | 4 | | 5 | | 6 | Row 0 |
|---|---|---|---|---|---|---|---|---|---|---|---|
| * | | * | | * | | * | | * | | * | Row 1 |
| | * | | * | | * | | * | | * | | Row 2 |
| | | * | | * | | * | | * | | | Row 3 |
| | | | * | | * | | * | | | | Row 4 |
| | | | | * | | * | | | | | Row 5 |
| | | | | | * | | | | | | Row 6 |

Example 1: $f(x) = x^2$

| 1 | | 2 | | 3 | | 4 | | 5 | | 6 | Row 0 |
|---|---|---|---|---|---|---|---|---|---|---|---|
| 1 | | 4 | | 9 | | 16 | | 25 | | 36 | Row 1 |
| | 3 | | 5 | | 7 | | 9 | | 11 | | Row 2 |
| | | 2 | | 2 | | 2 | | 2 | | | Row 3 |
| | | | 0 | | 0 | | 0 | | | | Row 4 |
| | | | | 0 | | 0 | | | | | Row 5 |
| | | | | | 0 | | | | | | Row 6 |

2. As a group determine the similarities in the tables. What general conclusions—for all polynomials, at least—would you conjecture?

## Part B

In part A, Row 2 contains numbers of the form $p(n+1)-p(n)$. Such a number can also be written as $\frac{(p(n+1)-p(n))}{1}$ or $\frac{(p(n+h)-p(n))}{h}$ where $h=1$. Thus, this number is an approximation to $p'(n)$. Similarly, Row 3 contains approximations of $p''$, etc. (Notice in the example that the values in Row 3 *are* the values of the second derivative of $y=x^2$.) We now want to see if we can "recover" a function if we know something about its "finite difference table"; that is, we want to anti-differentiate from a table. We will concentrate on determining quadratic functions.

1. Each person in the group should write in a different number in the right-most position in Row 1 and then complete the table so that the table corresponds to a table for a function as in part A. Now, try to determine the function.

| 1 | | 2 | | 3 | | 4 | | 5 | | 6 | Row 0 |
|---|---|---|---|---|---|---|---|---|---|---|---|
| * | | * | | * | | * | | * | | * | Row 1 |
| | * | | * | | * | | * | | 1 | | Row 2 |
| | | 2 | | 2 | | 2 | | 2 | | | Row 3 |

   As a group, decide what the first entry you chose for Row 1 does? Could you put more than one entry in Row 1 to start and then complete the table?

2. As individuals, take the table in 1 and put a different number in the right-most entry of Row 2 and choose a number for the right-most entry in Row 1 as before and complete the table. Can you write an algebraic expression for the function? As a group, determine what role an initial entry in Row 1 plays in the expression for the function.

### Problems

Consider a table for a quadratic function that begins as follows:

| 1 | | 2 | | 3 | | 4 | | 5 | | 6 | Row 0 |
|---|---|---|---|---|---|---|---|---|---|---|---|
| * | | * | | * | | * | | * | | * | Row 1 |
| | * | | * | | * | | * | | * | | Row 2 |
| | | 1 | | 1 | | 1 | | 1 | | | Row 3 |

Determine at least one quadratic function that has such a table. What role does the value in Row 3 play in the determination of the algebraic expression of the function?

## Using the derivative

Below are graphs of a function $f$ and its derivative $f'$.

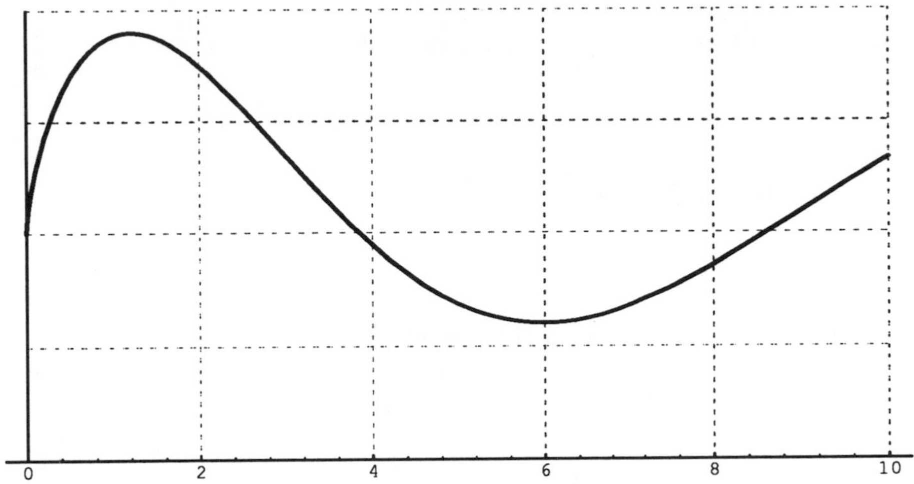


$y = f(x)$

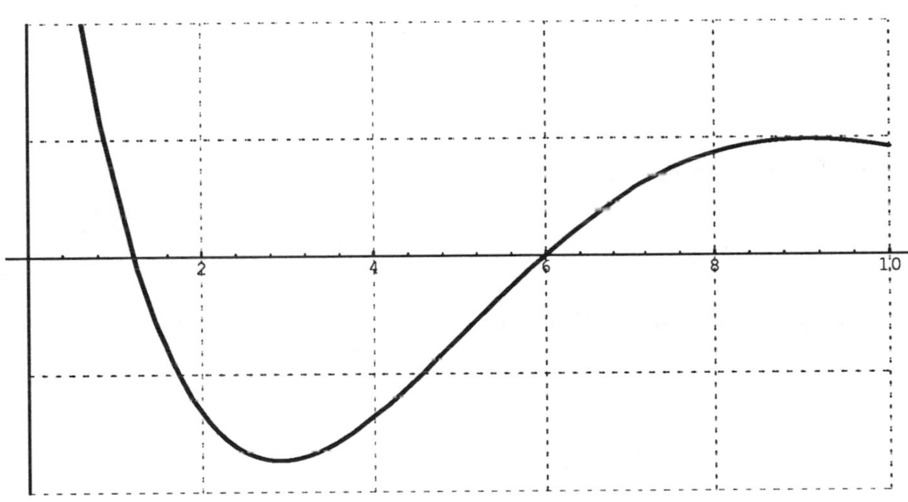


$y = f'(x)$

1. For what values of $x$ is $f$ increasing?

2. What is the behavior of $f'$ on those intervals?

3. For what values of $x$ is $f'$ negative?

4. What is the behavior of $f$ on those intervals?

5. For what values of $x$ is $f'$ increasing?

6. What is the behavior of $f$ on those intervals?

7. Where is the graph of $f$ concave down?

8. What is the behavior of $f'$ on those intervals?

9. Where does $f$ take on a local maximum value?

10. What can you say about $f'$ there?

## Gotcha

The following passage comes from the article "'I think I'll go down to Tully' and catch some speeders," in the Sunday, June 20, 1993, Syracuse *Herald American.*

> Trooper Jay Schwenk sat behind the pilot in the state police plane and surveyed the traffic 2,000 feet below.
>
> Click. Schwenk started the sky timer as a car zoomed over a white stripe painted on Interstate 81 in Tully. Click. Schwenk stopped the timer when the car hit a second stripe, a quarter mile from the first.
>
> "Eighty-three on the white one," Schwenk told the pilot, Sgt. Gerd Wolf. Schwenk timed the car again just to be sure. This time the driver was clocked at 82 mph.
>
> "I've got one coming up now going 82," Schwenk told patrol cars waiting alongside the highway just south of the Onondaga County line. "It'll be a white vehicle in the passing lane."
>
> Seconds later, a trooper pulled the car over.
>
> ...

As a group, answer the following.

1. How do you think that the car speed was computed?

2. Do you think that the driver had to be actually travelling the reported car speed at any instant?

3. Draw at least two different distance functions and their associated velocity functions for situations where the quarter mile would have been traversed in 10 seconds. What would have the trooper report? In each case is there a time when the velocity equals this reported speed?

**Problem**

Gas is flowing out of a pipe in such a way that it fills a one-gallon container in 10 seconds. Do you think it is possible that the rate at which the gas is flowing into the container is never 0.1 gallons per second? If so, construct an example of a function that gives the amount of gas in the container as a function of time where the rate of flow is never 0.1 gallons per second. Give a concrete scenario for your function. If you don't think this is possible, justify your conclusion.

## Animal growth rates

You are involved in a research project that involves working with a species of laboratory animal. If $W(t)$ is the weight (in ounces) of such an animal $t$ weeks after birth, then the growth of a healthy animal can be modeled by the differential equation

$$W'(t) = \frac{10}{W(t)} \quad \text{or} \quad \frac{dW}{dt} = \frac{10}{W}.$$

1. Describe in words what the differential equation says about the growth of a healthy animal.

Suppose you are responsible for an animal that weighs 5 ounces one week after it was born.

2. What does the model tell us about the rate of growth of the animal when it is one week old?

3. Give the equation of the line tangent to the graph of $W(t)$ at the point $(1,5)$.

4. Use the tangent line to approximate the weight of the animal 8 days after it was born.

5. Classify the graph of $W(t)$ for $t \geq 1$ as increasing and concave up, increasing and concave down, decreasing and concave up, or decreasing and concave down.

**Problem**

Suppose instead that the growth is modeled by

$$W'(t) = \frac{10}{W(t)^2} \quad \text{or} \quad \frac{dW}{dt} = \frac{10}{W^2}.$$

Answer 1–5 above.

## The product fund

Pat is employed by a financial services firm. One of her tasks is to execute transactions for her best client. The client is currently investing in the Drake mutual fund. The client has decided to purchase shares in the fund everyday instead of simply making a single large purchase. (This method of investing is sometimes called an "averaging" method since buying at several different times will mean that the price paid for the mutual fund will tend to be about the average price of the mutual fund over the time period. If the client purchased all her shares at once the client runs the risk of buying at the highest price which occurs during a period of time.)

1. The client owns 1000 shares of the Drake Fund at the start of business on Monday morning. The client calls Pat and wants to know the value of her Drake Fund investment. Pat looks at her workstation and sees that Drake Fund is currently trading at \$13.50 per share. What is the current value of the client's Drake Fund investment?

2. By the end of the trading day on Monday two things have happened: (1) the price of a share of the Drake Fund has fallen \$0.25, and (2) Pat has purchased 100 more shares of the Drake Fund for her client.

   (a) What is the value of the client's investment in the Drake Fund at the end of the day?

   (b) What was the change in the value of the investment during the day?

Investors are very interested in whether the value of their investment is growing (this makes them happy) or declining (this makes them unhappy). However, investors are also very interested in the rate of change of the value of their investment. An investment growing at the rate of \$10 per day is much better than an investment that is growing at the rate of \$10 per year.

3. Pat decides to call the value of the investment of her client in the Drake Fund $t$ days after the start of business on Monday morning $V(t)$. She will call the number of shares of the Drake Fund that her client owns $t$ days after the start of business on Monday morning $N(t)$. She will denote by $p(t)$ the price of a share of the Drake Fund $t$ days after the start of business on Monday morning.

   (a) At any time $t$ there is a formula that relates $V$, $N$, and $p$. Write this formula.

   (b) Write the information contained in 1 in terms of $V$, $N$, $p$, and $t$.

   (c) Write a sentence that describes what $V'(t)$ means in the context of this activity.

   (d) What are the appropriate units for $V'(t)$?

(e) Write a sentence that describes what $N'(t)$ means in the context of this activity.

(f) What are the appropriate units for $N'(t)$?

(g) Write a sentence that describes what $p'(t)$ means in the context of this activity.

(h) What are the appropriate units for $p'(t)$?

4. Pat's client wants information about $V'(t)$, the rate of change of her investment. Pat has information about $N'(t)$, the rate of change of the number of shares that her client owns, since she has her client's purchase orders. She can look in the newspaper or use her workstation and get information about $p'(t)$, the rate of change of the price of a share of the Drake Fund. So, like any problem solver, she has to figure out how to use what she knows to get what she wants.

We will try to figure out what $V'(t)$ is by breaking the problem into two separate parts. (A good strategy for problem solving is to break a big problem into smaller problems and then try to solve the smaller problems. This is part of the *top-down* approach to problem solving.)

We will break the value of the investment into two parts: the first part is the value from shares bought at times less than or equal to $t$, and the second part is the value from shares bought *after* time $t$. We will call the shares bought at times less than or equal to $t$ "old shares." We will call the shares bought after time $t$ "new shares." Any share is an old share or a new share but *not both*, so the value of the investment is the total of the value of the new shares and the value of the old shares.

We will estimate what happens to the value of the client's investment $dt$ days after $t$. Using our strategy from above we will use the two categories of shares, "old shares," which were owned by time $t$, and "new shares," which were purchased some time after time $t$ but no later than time $t + dt$.

(a) Write down an expression for the number of old shares using $N$.

(b) Write down an expression for the number of new shares using $N$.

You should have obtained the following results: the number of old shares is $N(t)$ and the number of new shares is $N(t + dt) - N(t)$.

(c) Write a formula or expression for the value of the old shares at time $t+dt$.

(d) Write a formula or expression for the change in the value of the old shares between time $t$ and time $t + dt$.

(e) Since we are interested in the rate of change, divide the change in value of the old shares by $dt$ and let $dt$ approach 0. You now have the rate of change in the value of the old shares at time $t$.

(f) Write a formula or expression for the value of the new shares at time $t + dt$.

(g) How many new shares had been purchased by time $t$?

(h) What was the value of the new shares at time $t$?

(i) Write an expression for the change in the value of the new shares between time $t$ and time $t + dt$.

(j) What mathematical property of the function $p$ will guarantee that $p(t + dt) \to p(t)$ as $dt \to 0$?

(k) Since we are interested in the rate of change, divide the change in value of the new shares by $dt$ and let $dt$ approach 0. You now have the rate of change in the value of the new shares at time $t$.

(l) Add the expressions you obtained in parts 4e and 4k to obtain an expression for the rate of change in the value of the investment at time $t$.

(m) Write a sentence that will tell Pat how to calculate the rate of change of the value of the investment based upon what she knows.

5. Summarize what you have learned about finding the derivative of any function $V$ which is written as the product of two other functions such as $N$ and $p$ in terms of the functions $N$ and $p$ and their derivatives.

## Problems

1. The price of a share is \$14 and is rising at the rate of \$0.10 per day. Pat's client has 700 shares of the mutual fund and has instructed Pat to buy shares at the rate of 100 per day. What is the rate of change of the value of the client's investment in the mutual fund?

2. Pat knows that the shares in the mutual fund can't change by more than \$1.50 each day. (The exchange will halt trading in the shares if this happens.) She also has received instructions from her client to buy between 100 and 200 shares in the mutual fund each day. Pat is trying to get some idea of the rate of change of the value of the investment in the Drake fund. At the beginning of the day her client has 800 shares and each share is currently worth \$13.

   (a) Draw a graph of the price of a share for a day that will force the exchange to halt trading.

   (b) Draw a graph of the price of a share for a day that will *not* force the exchange to halt trading.

   (c) Decide whether it is possible to draw a graph of the price of a share for a day which will *not* force the exchange to halt trading but will have at least one point on the graph where the derivative of the price is larger than \$1.50/day.

   (d) Write a brief paragraph explaining why or why not Pat can give her client upper and lower estimates for the rate of change of the value of her investment at the beginning of the day.

   (e) Write a brief paragraph explaining why or why not Pat can give her client upper and lower estimates for the value of her investment at the end of the day.

## Exchange rates and the quotient rule

Recall that

$$\frac{d}{dx}\frac{1}{f(x)} = -\frac{1}{f(x)^2}f'(x).$$

Here we will investigate this phenomenon in the context of currency exchange rates.

According to the *New York Times*, the exchange rate for the Japanese yen on Friday, July 16, 1993, was 107.55 yen per U.S. dollar. The *Times* also reports that on the previous Friday the exhange rate was 109.94 yen per dollar.

Let $f(x)$ denote the exchange rate for yen (measured in dollars). Let $g(x)$ denote the exchange rate for dollars (measured in yen). For purposes of this activity, we will assume that $g(x) = \frac{1}{f(x)}$. (Strictly speaking, this is not accurate. At any time, if you exchange yen for dollars and then dollars for yen (or *vice versa*), you will end up with less money than you started.)

1. (a) Assume that you are holding dollars and you want to shop using yen. Did your dollars gain or lose value this week?

   (b) Assume that you are holding yen and you want to shop using dollars. Did your yen gain or lose value this week?

2. What do your answers to 1a and 1b say about the signs of $f'(x)$ and $g'(x)$?

3. Units are very important in this situation.

   (a) What are the appropriate units for $f(x)$?

   (b) What are the appropriate units for $g(x)$?

4. Using the data above, estimate $f'$ on Friday. (Use weeks to measure time. What are the units for $f'$?)

5. Using the data above, estimate $g'$ on Friday. (What are the units for $g'$?)

6. Check the quotient rule. Do the units work out? Can you explain why the answers do not match exactly?

## Problems

1. Use your estimate from 4 to estimate the exchange rates ($f(x)$ and $g(x)$) on Thursday, July 15.

2. Using this data point, estimate $g'$ on Friday. How does it compare to what you found in 5?

# Using the product rule to get the chain rule

## Part A

When you learned operations such as multiplication you only learned results such as $5 \times 3 = 15$. If you needed to compute the product of $3 \times 5 \times 4$ you used the idea of *associativity.* You computed $3 \times 5 = 15$ and then multiplied $15 \times 4 = 60$. This could be denoted $(3 \times 5) \times 4$.

1. Since you know how to find the derivative of a product of two functions, use the idea of associativity to find the derivative of the product of three functions $f$, $g$, and $h$. Your answer should involve $f$, $f'$, $g$, $g'$, $h$, and $h'$.

2. Find a rule for the derivative of the product of four functions $f$, $g$, $h$, and $k$.

3. Use the product rule to find the derivative of the product of two copies of the same function $f$. (We will denote this product as $f^2$.)

4. Use your result from 1 to find the derivative of the product $f \times f \times f$ or $f^3$.

5. Use your result from 2 to find the derivative of the product $f \times f \times f \times f$ or $f^4$.

6. What do you think the derivative of $f^n$ is for any positive integer $n$?

## Part B

1. Use the quotient rule to find the derivative of $1/f$. Using the traditional exponential notation we will call this $f^{-1}$. *Note: This does* not *stand for the inverse of $f$.*

2. Use the product rule to find the derivative of $1/f \times 1/f$. (We will write $f^{-2}$ to denote $1/f \times 1/f$.)

3. Write $f^{-3}$ as $1/f \times 1/f \times 1/f$ and use your result from 1 of part A to find the derivative of $f^{-3}$.

4. Write $f^{-4}$ as $1/f \times 1/f \times 1/f \times 1/f$ and use your result from 1 of part A to find the derivative of $f^{-4}$.

5. What do you think the derivative of $f^n$ is for any negative integer $n$?

6. What do you think the derivative of $f^n$ is for any integer $n$?

**Problems**

1. Compute the derivative of $\sin^{12} x$.

2. Is there a function whose derivative is $\sin^{11} x \cos x$?

3. (a) Find the derivative of $e^{nx}$ for any integer $n$.
   (b) What do you think the derivative of $e^{kx}$ is for any number $k$?

## Magnification

The *image* of an interval $[a, b]$ under a map $f$ (this is denoted as $f([a, b])$) is the set of all values $y$ that are equal to $f(x)$ for some $x$ in $[a, b]$. In shorthand,

$$f([a,b]) = \{y : y = f(x) \text{ for } a \leq x \leq b\}.$$

For example if $f(x) = x/2$, then $f([-2,2]) = [-1,1]$; if $f(x) = x^2$, then $f([-2,2]) = [0,4]$.

For each of the following functions find the image of each interval, the length of the image, and the length of the image divided by the length of the original interval. What do you think the magnification due to $f$ is when $x = 3$ for each of these functions?

1. If $f(x) = 3x$

| $[a,b]$ | length | $f([a,b])$ | length | ratio |
|---|---|---|---|---|
| $[2,4]$ | $4-2=2$ | | | |
| $[2.5,3.5]$ | 1 | | | |
| $[2.9,3.1]$ | 0.2 | | | |
| $[2.99,3.01]$ | 0.02 | | | |
| $[2.999,3.001]$ | 0.002 | | | |

Magnification at $x = 3$ is ____________.

2. If $f(x) = x^2$

| $[a,b]$ | length | $f([a,b])$ | length | ratio |
|---|---|---|---|---|
| $[2,4]$ | 2 | | | |
| $[2.5,3.5]$ | 1 | | | |
| $[2.9,3.1]$ | 0.2 | | | |
| $[2.99,3.01]$ | 0.02 | | | |
| $[2.999,3.001]$ | 0.002 | | | |

Magnification at $x = 3$ is ____________.

3. If $f(x) = -x^3$

| $[a, b]$ | length | $f([a, b])$ | length | ratio |
|---|---|---|---|---|
| $[2, 4]$ | 2 | | | |
| $[2.5, 3.5]$ | 1 | | | |
| $[2.9, 3.1]$ | 0.2 | | | |
| $[2.99, 3.01]$ | 0.02 | | | |
| $[2.999, 3.001]$ | 0.002 | | | |

Magnification at $x = 3$ is ____________.

4. If $f(x) = 1/x$

| $[a, b]$ | length | $f([a, b])$ | length | ratio |
|---|---|---|---|---|
| $[2, 4]$ | 2 | | | |
| $[2.5, 3.5]$ | 1 | | | |
| $[2.9, 3.1]$ | 0.2 | | | |
| $[2.99, 3.01]$ | 0.02 | | | |
| $[2.999, 3.001]$ | 0.002 | | | |

Magnification at $x = 3$ is ____________.

5. What do you think the magnification means *mathematically*? Include any other conjectures you think are true about magnification.

# Chapter 4

# Integration

## Time and speed

A car moved along a straight road and its speed was continually increasing. Speedometer readings were recorded at two-second intervals and the results were as follows:

| Time | 0 | 2 | 4 | 6 | 8 | 10 |
|---|---|---|---|---|---|---|
| Speed | 30 | 36 | 38 | 40 | 44 | 50 |

The speeds are given in feet per second, the times in seconds.

1. From the above information, one cannot tell exactly how far the car went in the ten seconds. Explain why this is true.

2. In the first two-second interval, what is the minimum distance the car could have travelled? What is the maximum distance?

3. In the second two-second interval, (i.e. between $t = 2$ and $t = 4$) what is the minimum distance the car could have travelled? What is the maximum distance?

4. In the third two-second interval, what is the minimum distance the car could have travelled? What is the maximum distance?

5. In the fourth two-second interval, what is the minimum distance the car could have travelled? What is the maximum distance?

6. In the fifth two-second interval, what is the minimum distance the car could have travelled? What is the maximum distance?

7. During the entire ten-second interval, what is the minimum distance the car could have travelled? What is the maximum distance? (These distances are called the *lower estimate* and the *upper estimate* of the distance travelled, respectively.) Explain where you used the assumption that the car's speed was increasing. What assumption could replace it and still give the same answer? What would the ride be like under your assumption?

8. If you had to guess how far the car went in the ten-second interval, what would you guess? What is the maximum difference (error) between your guess and the actual distance?

**Problems**

Suppose the following additional speedometer readings become available at the missing one-second times:

| Time | 1 | 3 | 5 | 7 | 9 |
|---|---|---|---|---|---|
| Speed | 32 | 37 | 39 | 41 | 49 |

1. Is the second set of information consistent with the first? What would the speed at the time 1 second have to be for it to be inconsistent?

2. Redo Questions 7 and 8 in the light of this new information.

3. If speedometer readings became available for each tenth of a second, i.e. for $t = 0.1, 0.2, 0.3, \ldots, 9.8, 9.9$, then by how much would your upper estimate for the distance traveled exceed your lower estimate?

4. If speedometer readings were available for each hundredth of a second, by how much would your upper estimate for the distance traveled exceed your lower estimate?

5. Explain why you can calculate how far the car went to any desired accuracy if you have access to the speedometer readings at every instant during the ten-second interval.

## Oil flow

Earlier this week, an oil tanker collided with a Coast Guard cutter off the California coast. The disabled tanker is spilling oil from its damaged hull.

The rate of flow of oil into the Pacific Ocean off the California coast was measured at several different time intervals yesterday. The rates are listed in the table below.

| Time | Amount (100 gal/hour) |
|---|---|
| 9:00 | 4.0 |
| 10:00 | 4.0 |
| 11:00 | 3.8 |
| 12:00 | 3.6 |
| 1:00 | 3.0 |
| 2:00 | 2.0 |
| 3:00 | 0.6 |
| 4:00 | 0.3 |
| 5:00 | 0.1 |
| 6:00 | 0.1 |
| 7:00 | 0.0 |

1. Draw a graph of the rate at which oil was spilling into the ocean as a function of time.

2. Estimate the total amount of oil that spilled during the ten hour period covered by the table. Explain your method. That is, explain any assumptions you are making.

3. What can you say about the error involved in part 2?

4. What extra information would allow you to improve your estimate in part 2? Try to be as specific as you can.

## Can the car stop in time?

### Part A

A car is traveling at 60 feet per second when the driver spots a deer in the road 300 feet ahead and slams on the brakes. The following readings of the car's speed at various times are given in the table below. Time is measured in seconds (since the driver applied the brakes) and speeds are measured in feet per second.

| Time | 0 | 2 | 4 | 6 | 7 |
|---|---|---|---|---|---|
| Speed | 60 | 50 | 30 | 12 | 0 |

1. Give an estimate for the minimum distance the car will travel after the brakes are applied. Show your work. State any assumptions you are making.

2. Give an estimate for the maximum distance the car will travel after the brakes are applied. Show your work.

3. If the deer freezes and does not move, will the car hit the deer? Explain.

### Part B

If the car was traveling at 60 feet per second when the driver saw the deer and the car slowed down at the rate of 8 feet per second per second, would the car hit the deer? Explain.

## Fundamental theorem of calculus

In this activity we will be trying to find areas under curves. We will start with a simple example.

1. Sketch the curve $y = 2t + 3$.

2. Find the area under this curve between the lines $t = 1$, $t = 4$, and the $t$-axis using geometry.

3. Find the area under this curve between the lines $t = 1$, $t = x$, and the $t$-axis using geometry.

Your answer to question 3 should involve $x$, and we can think of this formula as representing a function which gives the area under the curve between $t = 1$ and $t = x$. We will call this area function $A(x)$.

4. Evaluate $A(4)$ and check that you get the same answer as you did to question 2.

Next we will look at the rate of change of A with respect to x.

5. Pick a value $x > 1$, and indicate the region on your graph from question 1 whose area corresponds to $A(x)$.

6. Pick a value of $h > 0$ and indicate the region on your graph from question 1 whose area corresponds to $A(x + h)$.

7. Use geometry to find an algebraic expression for the area of the region in your original graph which corresponds to $A(x+h) - A(x)$. (Your answer will involve both $x$ and $h$.)

8. Divide your answer to question 7 by $h$.

9. Look at your answer to question 8 and let $h \to 0$.

10. In the language of calculus describe what questions 7, 8, and 9 accomplished.

11. Look at your answer to question 9 and the original function defined in question 1. How are they related?

12. Describe in the language of calculus how the function $A$ you found in question 3 and the original function given in question 1 are related.

**Problem**

If the function you were given was not linear (such as $x^3 + 2$)—so that you could not find $A(x)$ by geometry—can you think of another way to find $A(x)$ that will work? Write a detailed description of your method, using the language and notation of calculus.

## Comparing integrals and series

1. Figure 1 (below) indicates a pattern for the creation of infinitely many boxes inside a triangle in the plane. If the areas of all the boxes were collected to form a series, think about whether or not the sum of the series would be the area of the triangle. Suppose your group was a group of attorneys and that your firm had to defend an answer to this question in a courtroom, with this figure, in evidence and where the jurors had not had calculus. You will need to consider arguments for both a yes answer and a no answer to this question. Without computing the sum of the series at this point, have your group prepare an intuitive argument why the answer might be yes as well as an intuitive argument why the answer might be no. Which argument does your group actually support at this time?

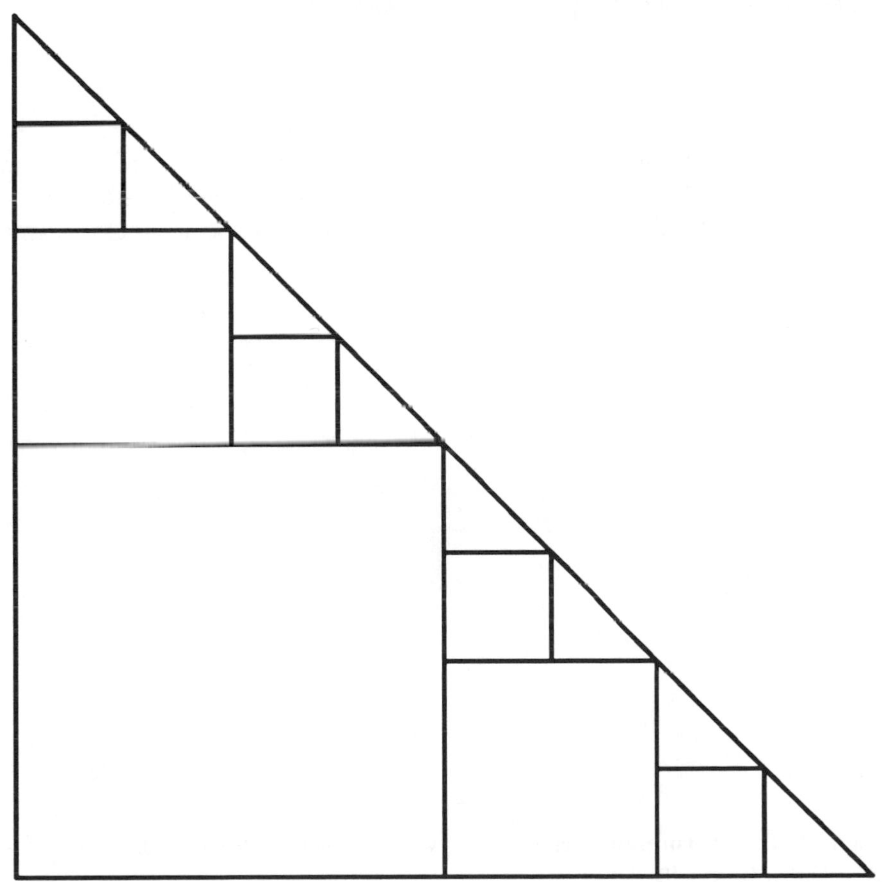

Figure 1

2. *Compute* the actual sum of the series from part 1. Does the sum of the series equal the area of the triangle? Does this provide conclusive evidence for either a yes answer or a no answer to the question raised in part 1?

3. *Represent* the area of the triangle as an integral.

4. What is the relationship between the value of the integral and the sum of the series? Does the series represent a Riemann sum?

5. Each member of your group should draw a (different) graph of a continuous, positive valued function, $y = f(x)$, over a closed interval in the plane. As a group, do you think that $\int f(x)\,dx$ can be represented as a series for any such function $f$? Support your conclusions.

## Problems

1. For $m > 0$, determine a series that gives the area under the curve $y = 1 - mx$ over the interval $[0, \frac{1}{m}]$.

2. Formulate a possible "real life" scenario around incremental depletion of a resource or commodity, such as a certain percent of the total resource or commodity was depleted each year, where the question as to whether or not such incremental depletion would exhaust the resource in the long run might be argued in a courtroom. Modify the arguments your group, or some group, gave in 1 of the activity to fit this scenario.

3. Using software (e.g., *Mathematica*), write a routine that "generates a series" solution to the area under a curve determined by the graph of a positive-valued, decreasing, concave down function over a closed interval. That is, find an algorithm that builds a partial sum approximation to the area. Discuss advantages and disadvantages of this algorithm over the right-hand endpoint algorithm.

## Graphical integration

1. Estimate the area of the region in the graph between $x = 1$, $x = 5$, $y = f(x)$, and the $x$-axis.

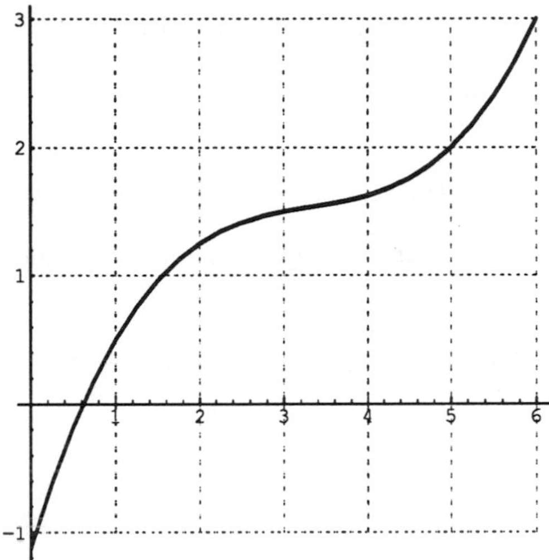


$y = f(x)$

2. Find the volume of the region formed by revolving the region described in 1 around the $x$-axis.

3. Find the area between $y = f(x)$, $y = 3$, $x = 1$, and $y = 1.5$.

4. Find the volume formed by revolving the region described in 3 around the $y$-axis.

**Problems**

1. Find the average value of the function $f(x)$ over the interval $[1, 5]$.

2. (a) Find the area of the region between $x = 1.5$, $y = f(x)$, $x = 3$, and the $x$-axis.

   (b) Express the area in 2a as a definite integral.

3. Find the volume of the solid formed by revolving the region described in 2a around the $x$-axis.

4. Find the area of the region between $y = 2.5$, $y = f(x)$, $y = 1.5$, and $x = 2$.

5. Find the volume obtained by revolving the region described in 4 around the $y$-axis.

6. How far is it along the curve $y = f(x)$ from $(1, 0.5)$ to $(3, 1.5)$?

# How big can an integral be?

## Part A

1. Plot the points

$$(0,1),\ (1,2),\ (2,3),\ (3,4),\ \text{and}\ (4,5).$$

2. Draw a straight line that passes through these points and is defined for all values in $[0,5]$.

3. Calculate the area under the graph you drew in 2, from $x = 0$ to $x = 5$.

4. Express your answer to 3 as an integral.

5. Draw a continuous curve that passes through the points and whose integral over $[0,5]$ is greater than 25.

6. Draw a continuous curve that passes through the points and whose integral is greater than 125.

7. Draw a continuous curve that passes through the points and whose integral is greater than 500.

8. Is there a positive number that is larger than the integral of every continuous curve that passes through the points? Explain.

9. Is there a continuous curve that passes through the points and whose integral is negative? Explain.

### Part B

In this part of the activity, let the vertical axis measure velocity with scale 1 unit = 10 mph and the horizontal axis measure time with scale 1 unit = 1 hr. Let all the graphs you drew in 2, 5, 6, and 7 represent the velocity of a car over a five-hour period.

1. Interpret the points given in 1 of Part A.

2. Interpret the curve you drew in 2 and the integral in 4 of Part A.

3. Interpret 6 of Part A.

4. Answer 8 of Part A in this context.

5. Answer 9 of Part A in this context.

### Part C

The curve you found in 2 (Part A) has a derivative equal to 1 for all $x$. In this section assume all the curves you drew in Part A have a derivative at every point and the value of the derivative at each point is greater than or equal to 0.

1. Answer 8 of Part A.

2. Answer 9 of Part A.

## Numerical integration

1. Use six rectangles of equal base and the midpoint method to estimate the area in the first quadrant under the curve $y = 4 - (x - 1)^2$ from $x = 0$ to $x = 3$.

2. Let us define the error associated with an estimate to be the difference between the estimated value of the area and the exact value. That is

$$\text{error} = \text{estimate} - \text{exact value}$$

Compute the exact value of the area in question 1, and the error associated with the estimate that you computed.

3. Now estimate the same area, this time using six trapezoids. Again, compute the error associated with the new estimate.

4. Compare the errors associated with the midpoint and the trapezoidal methods. Devise a way to combine the two computations that will produce a better estimate than either of these.

5. Compute the weighted average $\dfrac{(2\,\text{Mid} + \text{Trap})}{3}$. Compute the error associated with this estimate.

**Problems**

Repeat this activity, this time estimating the area under the curve $y = x + \sin x$ from $x = 0$ to $x = \pi$, and using four subintervals. Choose the best estimate. Discuss how this problem differs from the original activity.

## Verifying the parabolic rule

Let $f(x) = x - 1/x^2$. We will estimate the area under the curve from $x = 1$ to $x = 3$.

1. Draw a graph of the region.

2. Estimate the area using the midpoint method and one rectangle.

3. Estimate the area using one trapezoid.

4. Compute the weighted average $\dfrac{(2\,\text{Mid} + \text{Trap})}{3}$ (Simpson's rule, with $n = 2$).

5. The endpoints of the interval $[1, 3]$ and the midpoint correspond to the three points $(1, 0)$, $(2, 7/4)$, and $(3, 26/9)$ on the graph of $f(x)$. Verify that these three points also lie on the graph of the quadratic function
$$q(x) = \frac{11}{36}x^2 + \frac{30}{36}x - \frac{41}{36}.$$

6. Compute the area under the quadratic function in question 5, and compare your answer with the answer to 4.

7. On the same set of axes sketch the graphs of $f$ and $q$ between $x = 1$ and $x = 3$.

8. In your own words, describe what this activity illustrates.

## Finding the average rate of inflation

Consider the graph, which gives the rate of inflation (in percent per year) in a country with economic difficulties for the years indicated.

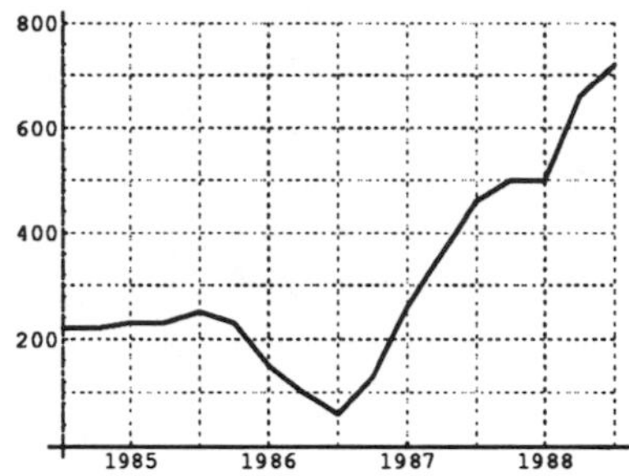


Inflation rate

1. Estimate the average rate of inflation over the time period shown.

2. Draw a horizontal line with height equal to the average rate of inflation on the graph. Relate the area under the horizontal line to the area under the graph of the inflation rate.

### Problems

1. What was the average rate of inflation for the year 1986?
2. How much money would it take at the end of 1988 to have the same buying power as \$100 at the beginning of 1985?
3. In a different country, the rate of inflation $t$ years after 1986 is 4% times $t$.
   (a) What is the average rate of inflation for this country from 1985 to 1993?
   (b) Draw a graph of the rate of inflation versus time and indicate on the graph what the average is.

# Cellular phones

A car is travelling on a straight road on a stretch that contains cities A, B, C, and D as illustrated below. The car is between cities A and D.

City A is 60 miles from City D, with City B 20 miles from cities A and C and City C 20 miles from city D. There are cellular phone receiving stations in each of the four cities. Each station has a range of ten miles.

1. Suppose that the car is travelling at a uniform rate of 55 miles per hour. What percentage of the time for the trip between cities A and D is spent within range of the station in City A? City B?

2. Suppose that the velocity of the car is

$$r(t) = \begin{cases} 60t, & t \leq 1 \\ 120 - 60t, & t \geq 1 \end{cases}$$

   where $t$ is measured in hours and $r(t)$ is measured in miles per hour. Also, suppose that at $t = 0$, the car is at City A. Now, what percentage of the time for the trip between cities A and D is spent within range of the station in City A? City B?

3. Do you think that it is possible to start driving from City A, stop driving at City D, and maintain the same percentage of time within range of each location? If so, draw a graph of the velocity of a car on such a trip. If not, explain why not.

## Problems

1. Given the velocity of the car as described in part 2, do you think that you could move the location of the stations to other locations in such a way that the car would spend the same amount of time within range of each station? If so, justify your answer. If not, explain why not.

2. Below is a graph of a velocity of a car as it travels between City A and City D. What percentage of the time is spent within range of City B?

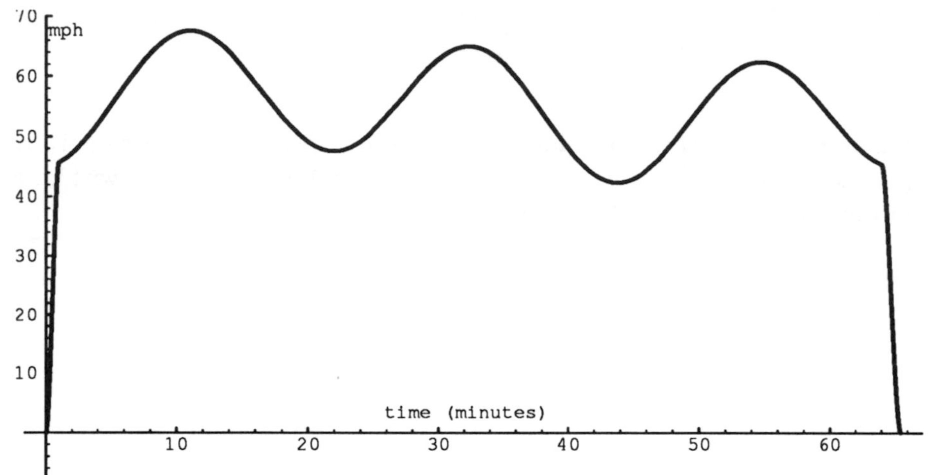


Velocity graph

## The shorter path

1. Find the distance from the origin to the point whose coordinates are $(1, 1)$.

2. Pat and Chris are hiking and they want to make camp at a campsite that is 1 mile east and 1 mile north from their current location. How far (as the crow flies) are they from the campsite? Write a brief explanation of how the answer to question 1 can be used to answer this problem.

3. When Pat and Chris look at their map they see that there is a lake between the campsite and their current location. If they follow the path $y = \sqrt{x}$ (the Radical Road) they will skirt the northern edge of the lake and if they follow the path $y = x^3$ (the Cubic Camp Trail) they will skirt the southern edge of the lake.

   (a) Draw a picture of the situation.

   (b) Explain why both paths will reach the campsite.

4. Deciding to estimate the length of each path, they do the following. Each path has a rest stop located at a point on the path which is $\frac{1}{2}$ mile east and some distance north of the starting point.

   (a) How far north is each rest stop from their starting position?

   (b) How far (as the crow flies) is each rest stop from their starting point?

   (c) How far (as the crow flies) is each rest stop from the camp site?

(d) Now give an estimate for the total distance along each path to the campsite. Is your estimate larger than the actual distance along the path (an upper estimate) or smaller than the actual distance along the path (a lower estimate) or neither? Explain your reasoning.

5. Is it possible for the hikers to hike less than 1.5 miles and get to the campsite using one of the paths? Write a brief explanation of your answer.

6. Which path gives the shorter route to the campsite? Explain your choice.

**Problems**

1. (a) If additional rest stops were added to each path at distances $\frac{1}{4}$ mile and $\frac{3}{4}$ mile east of the starting point, how far north are each of these rest stops from the starting point?

   (b) Use the additional rest stops to improve the estimates you found in 4d.

2. If rest stops are located every $dx$ miles east of the starting point write an expression for an estimate of the distance along each path.

3. Use algebra or calculus to express your answer to 2 as a Riemann sum.

4. Express the distance along each path as an integral.

5. What is the distance along each path between the starting point and the campsite?

# The River Sine

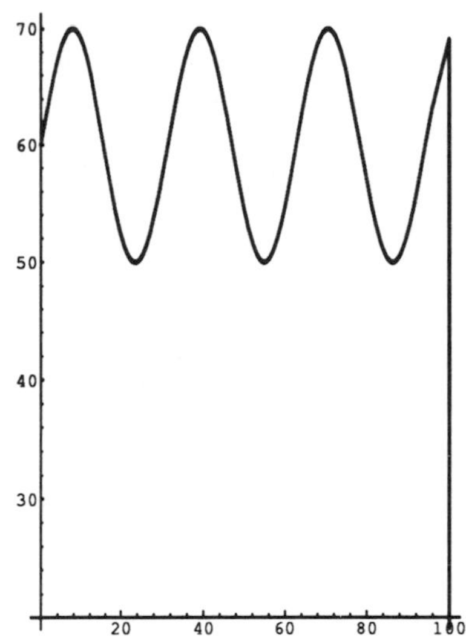


Your property

You own a plot of riverfront property which is pictured in the figure. Your property runs along the $x$-axis from $x = 0$ to $x = 100$ and is bounded by the lines $x = 0$, $x = 100$, and the River Sine whose equation is $y = 60 + 10\sin(x/5)$.

1. What is the area of the plot?

2. You have \$600 to spend, and you hire a gardening firm to install a fence along the riverside part of your property. The firm will charge you \$3 a foot for the fence including installation. If there is any money left over, the firm will fertilize the lot. They charge \$20 a bag for fertilizer. (This charge includes spreading the fertilizer.) One bag of fertilizer will cover 1,000 square feet.

   (a) Will there be any money left for fertilizer after the fence along the river is installed? Explain your answer.

(b) If there is money for fertilizer, about how much of your plot will be fertilized? Explain.

# Chapter 5

# Transcendental Functions

## Ferris wheel

We will investigate the motion of a person riding on a ferris wheel. The radius of the ferris wheel is 40 feet. The ferris wheel rotates in a counter-clockwise direction.

### Part A

We will start with the simple case in which the wheel rotates through an angle of one radian each second.

1. Set up a coordinate system with the origin at the center of the wheel. Let the point where the person is sitting be represented by the point whose coordinates are $(40, 0)$ to start.

   Figure 1: Wheel

   On your coordinate system, plot where the person will be 1 second later, $\pi/2$ seconds later, 2 seconds later, 3 seconds later, $\pi$ seconds later, $3\pi/2$ seconds later, $2\pi$ seconds later.

   What has happened in the first $2\pi$ seconds?

2. Now, on Figure 2, plot the horizontal position of the person for 10 seconds.

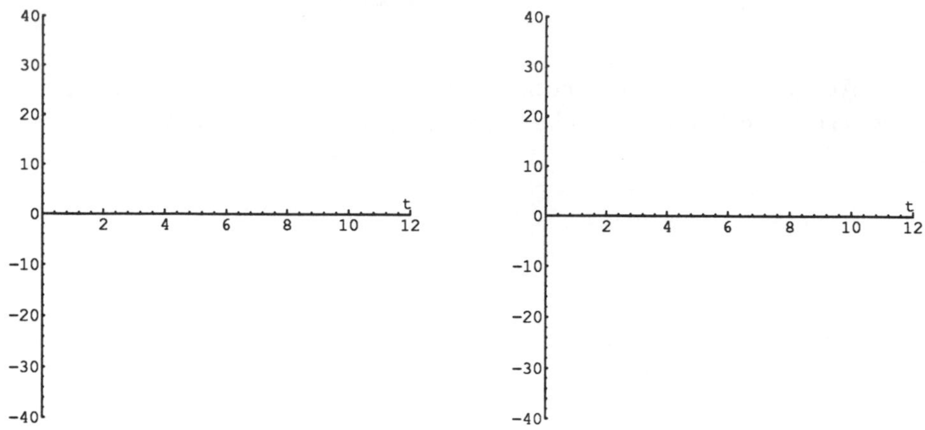


Figure 2: Horizontal

Figure 3: Vertical

3. On Figure 3, plot the vertical position of the person for 10 seconds.

4. Now go back and draw the horizontal and vertical positions on the same axes you used above, but assuming the wheel rotates at 2 radians per second.

5. Draw the graph of the horizontal position for 12 seconds if the wheel rotates at 1 radian per second for 8 seconds and then gradually increases its speed over the next 4 seconds to 2 radians per second and then continues to rotate at 2 radians per second.

6. Finally draw a graph of the horizontal position for what you think is a realistic ferris wheel and explain your graph.

## Part B

Now we will relate this work to functions we have already seen.

1. Draw a picture of the ferris wheel with the person somewhere in the first quadrant. (We will call this point $P$.) Drop a perpendicular line to the horizontal axis from the point where the person is sitting. The point where this line hits the horizontal axis we will call $Q$.

2. Look at the right triangle formed by the points $P$, $Q$, and the origin, $O$.

   Denote the angle formed by $Q\,O\,P$ as $\theta$ and remember that the length of $O\,P$ is 40 because the radius of the ferris wheel is 40 feet.

   (a) Using trigonometry write an expression for the length of $O\,Q$. ____________

   (b) Using trigonometry write an expression for the length of $P\,Q$. ____________

3. Suppose the wheel is rotating at 1 radian per second.

   (a) Find a rule for $\theta$ in terms of time. $\theta =$ ____________

   (b) Write a rule for the horizontal position of the person in terms of time. $x =$ ____________

   (c) Write a rule for the vertical position of the person in terms of time. $y =$____________

4. Suppose the wheel is rotating at 2 radians per second.

   (a) Find a rule for $\theta$ in terms of time. $\theta =$ ____________

   (b) Write a rule for the horizontal position of the person in terms of time. $x =$ ____________

   (c) Write a rule for the vertical position of the person in terms of time. $y =$____________

5. (a) Draw a graph of the horizontal *velocity* of the person if the wheel is rotating at 1 radian per second.

(b) Draw a graph of the vertical velocity of the person if the wheel is rotating at 1 radian per second.

## Why mathematicians use $e^x$

1. (a) Apply the definition of derivative to write an expression for the derivative of the function $f(x) = 2^x$ at the point whose first coordinate is $x$.

   (b) Using the properties of exponents, simplify your answer to 1a, and show that it can be expressed as

$$2^x \lim_{h\to 0} \frac{2^h - 1}{h}.$$

Notice that, if we can find the limit mentioned in 1b, we will know the derivative of $2^x$ at every point. But the limit in 1b is the slope of the tangent line to the graph of $y = 2^x$ at the point where $x = 0$.

2. Draw the graph of $y = 2^x$ with $x = 0$ in the center of your graph.

3. Draw the tangent line to the graph at the point $(0, 1)$ on the graph you drew in 2.

4. Now we will use the definition of derivative to find the slope of the tangent. On the graph you drew in 2 draw the secant line from $(0, 1)$ to the point on the graph of $y = 2^x$ whose first coordinate is 1.

   (a) Compute the slope of this secant line.

   (b) Is the slope of the secant line larger or smaller than the slope of the tangent line?

(c) What do the results of 4a and 4b tell you about the slope of the tangent line to the graph at $(0,1)$?

5. Next, on the graph you drew in 2, draw the secant line from $(0,1)$ to the point on the graph of $y = 2^x$ whose first coordinate is $-1$.

   (a) Compute the slope of this secant line.

   (b) Is the slope of the secant line larger or smaller than the slope of the tangent line?

   (c) What do the results of 5a and 5b tell you about the slope of the tangent line to the graph at $(0,1)$?

   (d) Using the results of 4 and 5, what can you now say about the slope of the tangent line to the graph at $(0,1)$?

6. Recall the definition of the derivative of the function $f(x) = 2^x$ at the point $x = 0$ is

$$f'(0) = \lim_{h\to 0} \frac{2^h - 1}{h}.$$

   (a) What value of $h$ was used for the secant line you worked with in 4?

   (b) What value of $h$ was used for the secant line you worked with in 5?

   (c) Rephrase your results of 5d in the language of derivatives.

7. Now we will fill in the tables below. (There is one table for positive values of $h$ and a second table for negative values of $h$.)

| | positive $h$ | | negative $h$ |
|---|---|---|---|
| $h$ | $\frac{f(0+h)-f(0)}{h}$ | $h$ | $\frac{f(0+h)-f(0)}{h}$ |
| 0.1 | | −0.1 | |
| 0.01 | | −0.01 | |
| 0.001 | | −0.001 | |
| 0.0001 | | −0.0001 | |

(a) Fill in the table for $h = 0.1$ and $h = -0.1$. What can you now say about the derivative of $y = 2^x$ at $x = 0$?

(b) Fill in the table for $h = 0.01$ and $h = -0.01$. What can you now say about the derivative of $y = 2^x$ at $x = 0$?

(c) Fill in the table for $h = 0.001$ and $h = -0.001$. What can you now say about the derivative of $y = 2^x$ at $x = 0$?

(d) Fill in the table for $h = 0.0001$ and $h = -0.0001$. What can you now say about the derivative of $y = 2^x$ at $x = 0$?

(e) How many decimal places do you now know of the value of $f'(0)$? Explain.

(f) What do you know about the derivative of $2^x$?

8. (a) What can you say about the slopes of the secant lines that correspond to negative $h$ as $h$ approaches 0?

(b) What can you say about the slopes of the secant lines that correspond to positive $h$ as $h$ approaches 0?

(c) Relate what you found in 8a and 8b to the concavity of the graph of $y = 2^x$.

(d) Summarize how you can estimate the value of the derivative at a point if you know the concavity of the graph in an open interval that contains that point.

**Problems**

1. Repeat parts 1, 2, 3, 5, 6, and 8 for the function $f(x) = 3^x$.
2. Repeat parts 1, 2, 3, 5, 6, and 8 for the function $f(x) = 4^x$.
3. If there is a number $a$ with the property that the derivative of $a^x$ at $x = 0$ is 1 what can you say about the number $a$? (Based on the results of the activity and problem 1 and problem 2.)
4. Decide whether the number $a$ described in problem 3 is larger than 2.5. Explain your reasoning.
5. Decide whether the number $a$ described in problem 3 is larger than 2.75. Explain your reasoning.

## Exponential differences

1. Each member of the group should choose a different exponential function with integer bases (e.g., $y = 3^x$, $y = 4^x$, etc.) and construct a finite difference table as is illustrated in the example below. That is, if your function is $y = f(x)$, in Row 1 compute $f(1)$, $f(2)$, $f(3)$, $f(4)$, $f(5)$, $f(6)$. In Row 2, over the star "*" compute $f(2) - f(1))$, $(f(3) - f(2))$, $(f(4) - f(3))$, $(f(5) - f(4))$, and $(f(6) - f(5))$. In Row 3 and in subsequent rows, over the stars, compute the difference between the entry in the row directly above the star to the right and the entry in the row directly above the star to the left.

| 1 | | 2 | | 3 | | 4 | | 5 | | 6 | Row 0 |
|---|---|---|---|---|---|---|---|---|---|---|---|
| * | | * | | * | | * | | * | | * | Row 1 |
| | * | | * | | * | | * | | * | | Row 2 |
| | | * | | * | | * | | * | | | Row 3 |
| | | | * | | * | | * | | | | Row 4 |
| | | | | * | | * | | | | | Row 5 |
| | | | | | * | | | | | | Row 6 |

Example: $f(x) = 2^x$

| 1 | | 2 | | 3 | | 4 | | 5 | | 6 | Row 0 |
|---|---|---|---|---|---|---|---|---|---|---|---|
| 2 | | 4 | | 8 | | 16 | | 32 | | 64 | Row 1 |
| | 2 | | 4 | | 8 | | 16 | | 32 | | Row 2 |
| | | 2 | | 4 | | 8 | | 16 | | | Row 3 |
| | | | 2 | | 4 | | 8 | | | | Row 4 |
| | | | | 2 | | 4 | | | | | Row 5 |
| | | | | | 2 | | | | | | Row 6 |

As a group, determine any similarities in the patterns that develop. Can you describe what happens in general? (That is, what happens with a general exponential function $y = a^x$?)

2. If we combine the process of taking differences with division by entries in the previous row, the rows below row 1 become constant. By what entries in the previous row should you divide? How does the constant relate to the base of your exponential function?

## Inverse functions and derivatives

The inverse of a function $f$, if it exists, is another function, $g$, that reverses the action of $f$. For example, if the function $f$ gives the position, $s$, of an object as a function of time, $t$, then the inverse function, $g$, gives the time as a function of the position. That is, given any position, $s$, the function g gives the time, $t$, when the object is at position $s$.

In this activity, we will investigate inverse functions and their derivatives.

1. The function, $f$, whose graph is below gives the number of German marks one could obtain for $x$ U.S. dollars on October 1, 1993.

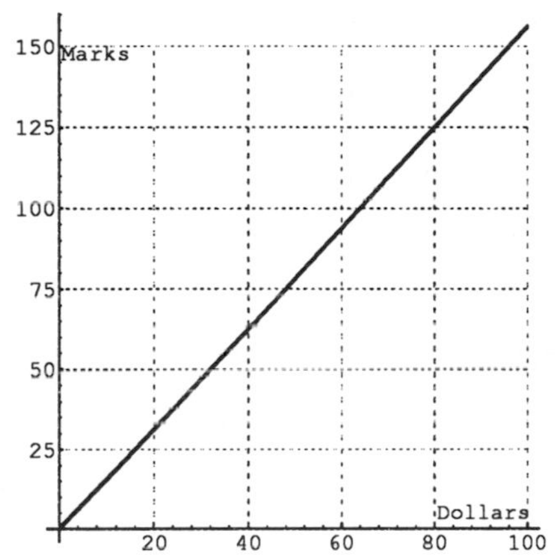


Dollars to Marks

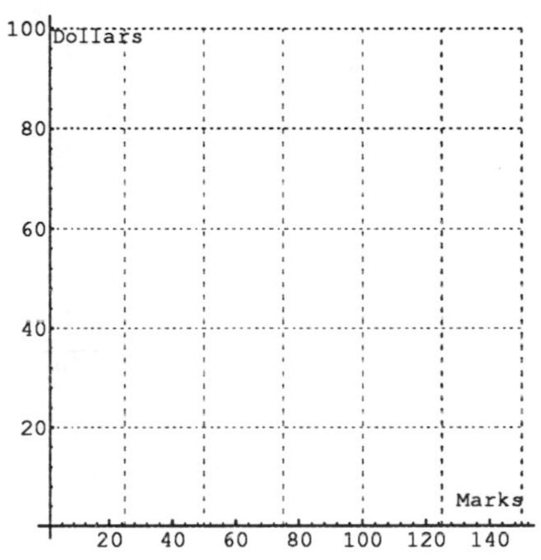


Marks to Dollars

(a) If I held \$20 US on October 1, how many German marks could I obtain?

(b) If I held U.S. dollars on October 1 and wanted to obtain German marks, what exchange rate was I dealing with? Explain how the exchange rate can be read from the original graph.

(c) If I held 40 German marks on October 1, how many U.S. dollars could I obtain?

(d) On the second set of axes above, sketch the graph of the function, $g$, that gives the number of U.S. dollars one can obtain for $x$ German marks. That is, sketch the graph of the function, $g$, that is the inverse of the given function, $f$.

(e) If I held German marks on October 1 and wanted to obtain U.S. dollars, what exchange rate was I dealing with? Explain how the exchange rate can be read from the graph you just sketched.

(f) How are the two exchange rates you found related?

(g) Describe the relationship between the two exchange rates using the language and notation of calculus.

2. An athlete, Jane, running a five-mile course has her coach plot her position as a function of time. The position is the distance, in miles, that she has run from the start of the course, and time is measured in minutes from the time she started her run. Call this function $f$. The graph is shown below.

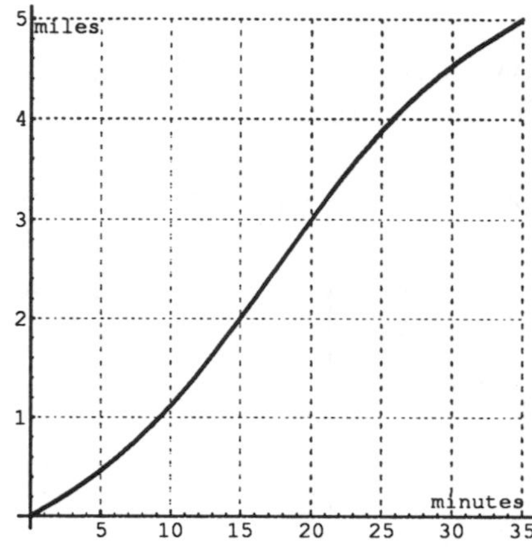


Miles vs. Minutes

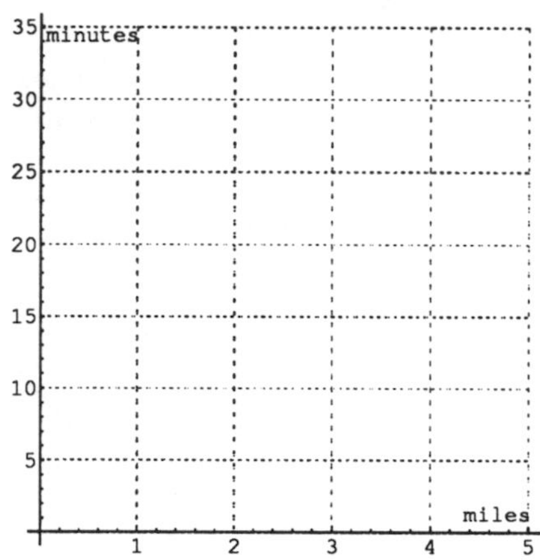


Minutes vs. Miles

(a) How long did it take Jane to run the first mile? the first two miles? the first three miles? all five miles?

(b) How fast, in miles per minute, was she running after ten minutes? Explain how to read this value from the original graph. How can this speed be described precisely, using the language and notation of calculus?

(c) Most runners, for practical purposes, measure their speed in minutes per mile and think of a course in terms of miles travelled so far and how long it took them to reach each milepost. On the second set of axes above, make a plot of Jane's time as a function of the distance travelled so far. Call this function $g$. It is the inverse of the original function, $f$.

(d) How fast, in minutes per mile, was Jane running at the end of the second mile? Explain how to read this value from the graph you just drew. How can this speed be described precisely, using the language and notation of calculus?

(e) Describe carefully, using the language and notation of calculus, how the two rates you found above are related.

## Fitting exponential curves

Consider the following data.

| $x$ | $y$ |
|---|---|
| 0 | 3 |
| 1 | 3.66 |
| 2 | 4.48 |
| 3 | 5.47 |
| 5 | 8.15 |
| 6 | 9.96 |
| 8 | 14.86 |
| 10 | 22.17 |
| 12 | 33.07 |
| 15 | 60.27 |

1. Sketch these points on a set of axes, and decide whether the function is increasing at a constant rate, increasing and concave up, or increasing and concave down. Explain you choice.

2. Calculate the natural logarithm of the $y$ values. Now on a new set of axes plot $\ln y$ versus $x$. What do you notice?

3. If a function has the form $y = ce^{bx}$, we say it is an exponential function. Take the natural logarithm of both sides of the above equation to show that $\ln y = \ln c + bx$.

4. If we let $Y = \ln y$ and $a = \ln c$ in the above we have that $Y = a + bx$, which says that $Y$ is a linear function of $x$. That is, if we plot $Y$ versus $x$ we should get a straight line. Use your ruler to draw a straight line through the points in your graph from part 2.

5. Estimate the slope and the intercept for the straight line you drew in part 4.

6. Using your answers from part 5 you now have estimates for $a$ and $b$. This will also allow you to find $c$ since $a = \ln c$. (Do you see why?) Now use these values to find an exponential function that fits the original data.

To summarize, we note that if we think a table of data might have come from an exponential function, we can plot the logarithms of the $y$ values versus the $x$ values and see if it looks like a straight line. We also note that the straight line we get also allows us to estimate the parameters $c$ and $b$.

**Problems**

The following gives the population of Mexico for the years 1980 to 1986.

| Year | Population (in millions) |
|---|---|
| 1980 | 67.38 |
| 1981 | 69.13 |
| 1982 | 70.93 |
| 1983 | 72.77 |
| 1984 | 74.66 |
| 1985 | 76.60 |
| 1986 | 78.59 |

It has been claimed that the population of Mexico grew exponentially during this period. What do you think? If you decide the answer is yes, find the exponential function which fits this data. What was the growth rate for the population during this time period? What is your estimate for the population in 1990? 1995?

## Log-log plots

Consider the following table of data.

| $x$ | $y$ |
|---|---|
| 1 | 3 |
| 2 | 8.49 |
| 3 | 15.59 |
| 5 | 33.54 |
| 6 | 44.09 |
| 8 | 67.88 |
| 10 | 94.87 |
| 12 | 124.70 |
| 15 | 174.28 |

1. Sketch these points on a set of axes and decide whether the function is increasing at a constant rate, increasing and concave up, or increasing and concave down. Explain your choice.

2. Calculate the natural logarithm of the $x$ and $y$ values. On a new set of axes plot $\ln y$ versus $\ln x$. What do you observe?

3. If a function has the form $y = cx^b$, $b > 0$, we say it is a power function. Take the natural logarithm of both sides of the above equation to show that $\ln y = \ln c + b \ln x$.

4. If we let $Y = \ln y$, $A = \ln c$, and $X = \ln x$, we have that $Y = A + bX$, which says that $Y$ is a linear function of $X$ with slope = ____________ and intercept = ____________. That is, if we plot $Y$ versus $X$ we should get a straight line. Use your ruler to draw a straight line through the points in your graph from part 2.

5. Estimate the slope and the intercept for the straight line you drew in part 4.

6. Using your answers from part 5 you now have estimates for $A$ and $b$. Use these estimates to find $c$. Now you have an approximation of the power function which fits the original data.

7. Use your approximation from part 6 to calculate approximate $y$-values for all the $x$-values in the table of data. Compare the computed $y$-values with the given $y$-values. How good is your approximation?

8. Write a brief summary of how we might test whether data may have come from a power function. Also indicate how you might estimate the power function.

**Problem**

Fit a power curve to the following data:

| $x$ | $y$ |
|---|---|
| 2 | 1.13 |
| 3 | 3.12 |
| 5 | 11.18 |
| 7 | 25.93 |
| 8 | 36.20 |
| 10 | 63.25 |
| 12 | 99.77 |
| 14 | 146.67 |

## Using scales

Beneath are three sets of coordinate axes. The first is a normal cartesian or $x$–$y$ coordinate system. The second set is called *semi-log* since the vertical scale is logarithmic. (So in normal coordinates you would be plotting points whose form was $(x, \ln y)$ instead of $(x, y)$.) The third is called *log-log* since both the vertical and horizontal scales are logarithmic. (This would be like plotting points of the form $(\ln x, \ln y)$ instead of $(x, y)$.)

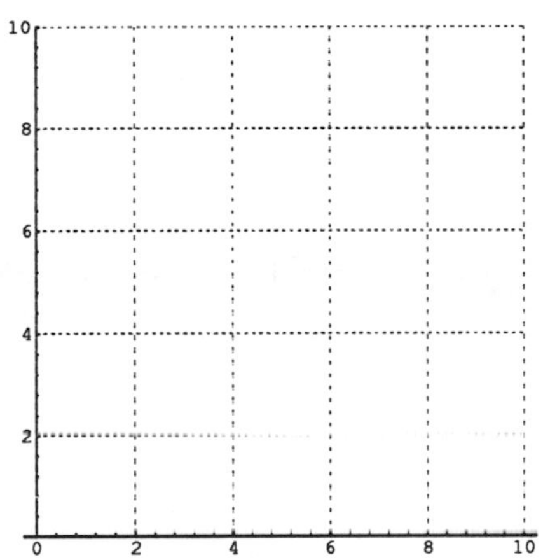


Cartesian

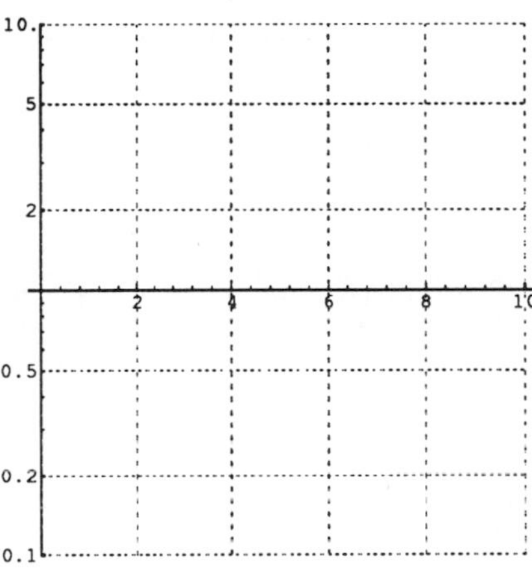


Semi-log

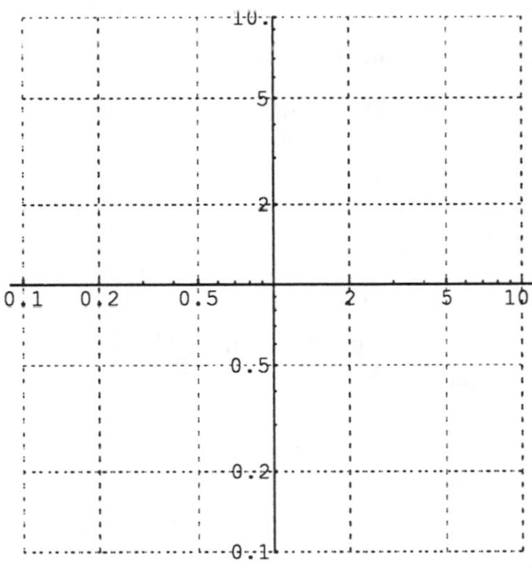


Log-log

Plot each of the relations on *all three axes.*

$$y = x + 2 \tag{5.1}$$
$$y = 2e^{.2x} \tag{5.2}$$
$$y = 3x^{.5} \tag{5.3}$$

1. There should be at least one set of axes where each relation looks linear. Identify each relation and the scale that makes it linear. Explain why that scale makes the graph linear.

2. Summarize your findings in a brief paragraph.

## Problems

1. You expect some data to follow the model $y = Cx^{\alpha}$.
   (a) What scale should you use to plot the data?
   (b) Explain how would you estimate $\alpha$.
   (c) Explain how you would estimate $C$.

2. (a) Look at the census data for the US from 1790 to 1860. Find the best model to fit the data.
   (b) Use your model to predict the population in 1870 and 1990.
   (c) Check with the actual population of the US in 1870 and 1990. Explain any discrepancies.

# Chapter 6

# Differential Equations

## Direction fields

This activity details a geometric method of solution for first-order differential equations. This method is called the method of direction fields and is based on the fact that a differential equation (DE) gives a direction at each point in the plane. For example, the DE $\frac{dy}{dx} = x$ says the slope of the curve $y(x)$ at the point $(x, y)$ in the $x$–$y$ plane equals the $x$ coordinate. So the slope of $y$ at the point $(1, 0)$ equals 1. In fact, the slope at any point $(1, y)$ equals 1.

We can use this information to approximate the solution curves to the given DE. We see from the above discussion that along the vertical line $x = 1$ the slope has the value 1. This line is called an *isocline.* For this differential equation, all the isoclines (curves of same slope) are vertical lines.

1. For this differential equation, find the isocline where the slope is $-1$.

2. For this differential equation, find the isocline where the slope is 0.5.

Suppose now that $\frac{dy}{dx} = x + y$.

3. Show that the isoclines of this DE are also straight lines. Find the lines where the slope is 1, $-1$, and 0.5.

Consider the original DE $\frac{dy}{dx} = x$.

4. On an axis draw the vertical line $x = 1$ and put a small dashed line of slope 1 at a few points on that to line indicate that the solution curve has slope 1 for all points on this line. What kind of dashed lines will you place on the vertical line $x = -1$? How about $x = 0.5$?

5. Continue in this fashion until you can estimate what the solution curves look like. What kinds of assumptions are you making here?

6. Find the solution that passes through the point $(0, 1)$.

### Problems

1. Use this method to estimate the shape of the solution to the DE $\dfrac{dy}{dx} = x + y$ that passes through the point $(0, 1)$. What happens to this solution as $x$ gets big?

2. Let $\dfrac{dy}{dx} = \dfrac{y}{x^2}$, $x \neq 0$. What are the isoclines for this DE? Sketch a solution curve that passes through the point $(1, 1)$.

## Using direction fields

Consider the differential equation described by

$$\frac{dy}{dt} = y\,(100 - y).$$

1. Find any constant solutions to the differential equation.

2. Draw some direction vectors for the differential equation.

3. Sketch the solution to the differential equation that passes through $y = 10$ when $t = 0$.

4. Write a short paragraph that describes the solution you found in part 3.

5. What will happen to the solution you found in part 3 as $t \to \infty$?

### Problems

1. Sketch the solution to the differential equation that passes through $y = 150$ when $t = 0$.

2. Write a short paragraph that describes the solution you found in 1.

3. What will happen to the solution you found in 1 as $t \to \infty$?

4. Try to classify *all* solutions to the equation by their behavior as $t \to \infty$.

## Drawing solution curves

For each of the direction fields given below:

1. Draw a solution curve through the point marked on the direction field.
2. Draw a solution that passes through $y = 1$ when $t = 0$.
3. Draw a solution that passes through $y = 3$ when $t = 2$.
4. List any constant solutions.

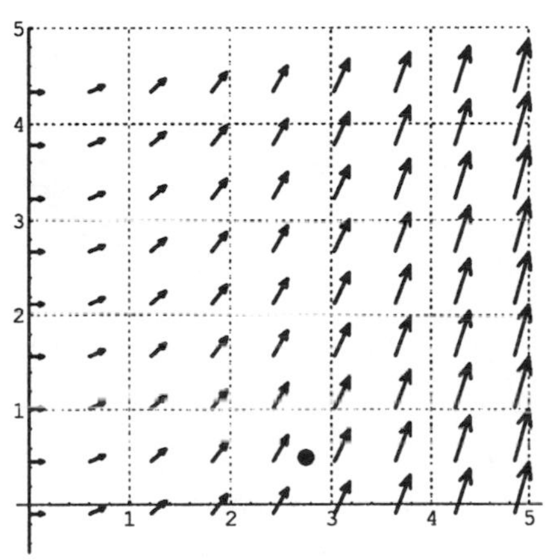

A

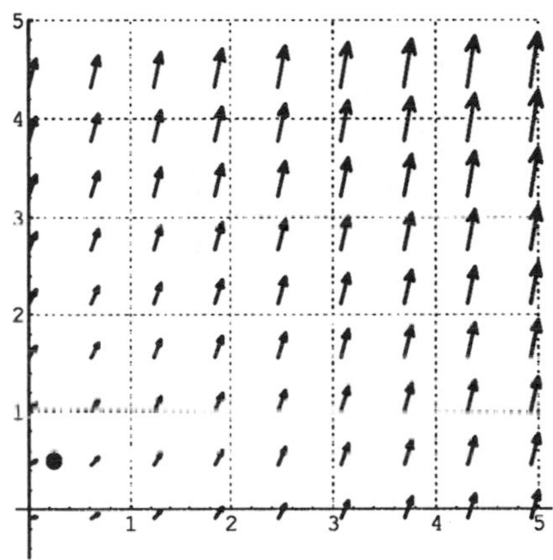

B

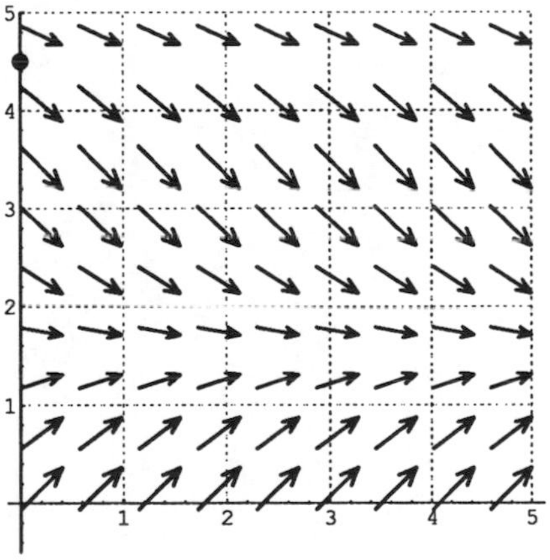

C

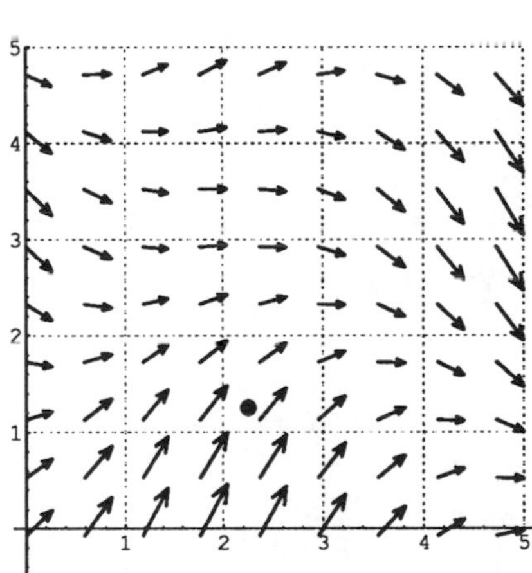

D

## The hot potato

Newton's law of cooling states that a body cools at a rate directly proportional to the difference between the body's temperature and the temperature of the surrounding medium.

This law can be written as

$$\frac{dT}{dt} = k(T - A)$$

where $T(t)$ is the temperature of the body at time $t$, $A$ is the temperature of the surrounding medium (the *ambient temperature*, which we assume to be constant), and $k$ is the constant of proportionality.

A hot potato of 300° F is placed on the kitchen table in a room kept at 72° F. The potato begins to cool according to Newton's law of cooling. Let us assume that $k = -0.12$, so the following differential equation governs this situation:

$$\frac{dT}{dt} = -0.12(T - 72) \qquad (*)$$

where $T(t)$ is the temperature of the potato at time $t$, and in particular $T(0) = 300°$ F, and time is measured in minutes.

1. If we substitute $t = 0$ into (*) we get $\frac{dT}{dt} = -0.12(T(0) - 72) = -0.12(300 - 72) = -27.36$ degrees per minute. What does this tell us about how fast the potato is cooling when it is first put on the kitchen table?

2. Estimate the temperature of the potato after it has been on the table for one minute.

3. Estimate the rate of change of temperature after one minute.

4. Use the information from parts 2 and 3 to estimate the temperature of the potato after it has been on the table for two minutes.

5. Estimate the temperature of the potato after it has been on the table for five minutes.

6. Try to find a step-by-step method to find the temperature of the potato at any time knowing its temperature at a particular time and knowing that it satisfies (*) all the time.

## Problems

1. How accurate do you think your method in part 6 is? What could you do to improve the accuracy of your method?

2. Check that $T(t) = 72 + 228e^{-.12t}$ is a solution to (*) satisfying $T(0) = 300°$. Use this rule to check the accuracy of your answers to the above parts.

3. Sketch a graph of temperature $T$ versus time $t$. What does the concavity of this graph say about how an object cools if the object obeys Newton's law of cooling? Physically does this make sense?

## Spread of a rumor: discrete logistic growth

A rumor spreads at a rate that is proportional to the product of the number of people in a population who have already heard the rumor (hearers) and the number of people who have not heard the rumor (ignorants). Using a discrete model for the spread of the rumor involves choosing a unit of time, $\Delta t$, and estimating how many additional people hear the rumor (converts) in each time period.

1. Using a discrete model for the spread of the rumor, fill in the table on the next page. (The numbers in parentheses assume rounding to the nearest integer at every step.)

$$\begin{aligned} \text{Population} &= 1000 \\ y &= \text{Number of hearers} \\ 1000 - y &= \text{Number of ignorants} \\ \text{Model: } \frac{\Delta y}{\Delta t} &= 0.001y(1000 - y) \\ \text{or } \Delta y &= 0.001y(1000 - y)\,\Delta t \\ \text{Step size, } \Delta t &= 1 \text{ day} \\ \text{so, } \Delta y &= 0.001y(1000 - y) \end{aligned}$$

2. Plot the number of hearers, $y$, as a function of the number of days since the start of the rumor.

3. Explain in words what the graph and the table indicate about how the spread of the rumor takes place.

| Day | # hearers($y$) | # ignorants ($1000 - y$) | # converts ($\Delta y$) |
|---|---|---|---|
| 0 | 5 | 995 | 4.975 (5) |
| 1 | 9.975 (10) | 990.025 (990) | 9.875 (10) |
| 2 | | | |
| 3 | | | |
| 4 | | | |
| 5 | | | |
| 6 | | | |
| 7 | | | |
| 8 | | | |
| 9 | | | |
| 10 | | | |
| 11 | | | |
| 12 | | | |
| 13 | | | |
| 14 | | | |

## Population

1. The birth rate in a state is 2% per year and the death rate is 1.3% per year. The population of the state is now 8,000,000.

   (a) At what rate are babies being born in the state now? Be sure to include units of measure in your answer.

   (b) At what rate are people dying in the state now?

   (c) Write a differential equation that the population of the state satisfies. Be sure to define your terms.

   (d) Solve the differential equation you wrote in part 1c

   (e) In how many years will the population reach 10,000,000?

   (f) Does the population have a steady state? Explain.

2. We will now try a more realistic model for the population. The birth rate and death rate are the same as in 1 and the population now is 8,000,000; however people are moving out of the state at a constant rate of 50,000 people per year.

(a) Write a differential equation for the population and solve it.

(b) Will the population ever reach 10,000,000? If so, when? If not, why not?

## Problems

1. For the situation described in 2:

   (a) Will the population ever reach 6,000,000? If so, when? If not, why not?

   (b) Does the population have a steady state? Explain.

   (c) At what constant rate will people have to leave the state in order for the state to have a constant population?

## Save the perch

Happy Valley Pond is currently populated by yellow perch. The pond is fed by two springs: spring A contributes 50 gallons of water per hour during the dry season and 80 gallons of water per hour during the rainy season. Spring B contributes 60 gallons of water per hour during the dry season and 75 gallons of water per hour during the rainy season. During the dry season an average of 110 gallons of water per hour evaporate from the pond and an average of 90 gallons per hour of water evaporate during the rainy season. There is a small spillover dam at one end of the pond and any overflow will go over the dam into Bubbling Brook. When the pond is full (i.e., the water level is the same height as the top of the spillover dam) it contains 475,000 gallons of water.

Spring B has become contaminated with salt and is now 10% salt. (This means that 10% of a gallon of water from Spring B is salt.) The yellow perch will start to die if the concentration of salt in the pond rises to 1%. Assume that the salt will not evaporate but will mix thoroughly with the water in the pond. There was no salt in the pond before the contamination of spring B. Your group has been called upon by The Happy Valley Bureau of Fisheries to try and save the perch.

### Part A

Let $t = 0$ hours correspond to the time when Spring B became contaminated. Assume it is the dry season and that at time $t = 0$ the pond contains 400,000 gallons.

1. Let $P(t)$ be the amount of water in the pond at time $t$. What will the change in the amount of water in the pond be during the time interval $t$ to $t + \Delta t$ for some (small) positive number $\Delta t$?

2. Using 1, find a differential equation for $P$.

3. Solve the differential equation and decide if the pond will fill up. If so how long after $t = 0$ until the pond is full?

### Part B

Use the same assumptions as Part A.

1. Let $S(t)$ be the amount of salt in the pond at time $t$. What will the change in the amount of salt in the pond be during the time interval $t$ to $t + \Delta t$ for some (small) positive number $\Delta t$?

2. Using 1, find a differential equation for $S$.

3. Solve the differential equation and decide what will happen to the amount of salt in the long run.

4. Draw a graph of the amount of salt in the pond versus time for the next three months.

5. How much salt will there be in the pond in the long run?

6. Do the fish die? If so when do they start to die?

## Part C

Answer the questions in Part B if the pond was full at time $t = 0$ and it was the dry season.

**Problem**

Answer the questions in Part B if the pond was full at time $t = 0$ and the contamination of Spring B occurred during the rainy season.

# Chapter 7

# Series

## Convergence

Decide whether or not each of the following statements is true *all the time.* If you think the statement is true write an explanation or a proof. If you decide it is not true give an example of a function that shows the statement is not true.

1. If the derivative of $f$ is positive for all $x > 0$, then $f(x) \to \infty$ as $x \to \infty$.

2. If $f'(x) \to 0$ as $x \to \infty$ then $f(x)$ converges to some *finite* number as $x \to \infty$.

## Investigating series

In this activity, you will experiment with some infinite sequences and their limits. Starting with a given sequence of numbers, $\{b_1, b_2, \ldots\}$, you will construct a new sequence $\{a_1, a_2, \ldots\}$ as follows:

$$\begin{aligned} a_1 &= b_1 \\ a_2 &= b_2 - b_1 \\ a_3 &= b_3 - b_2 \\ &\vdots \\ a_n &= b_n - b_{n-1} \\ &\vdots \end{aligned}$$

Starting with the following sequence as $\{b_n\}$:

$$\frac{2}{1}, \frac{8}{3}, \frac{26}{9}, \frac{80}{27}, \frac{242}{81}, \frac{728}{243}, \frac{2186}{729}, \ldots$$

1. Compute the first six elements of the sequence $\{a_n\}$.

2. Graph $\{a_n\}$ versus $n$ and $\{b_n\}$ versus $n$ on the same set of coordinate axes. Plot at least the first six values for each sequence. Visually determine the limit of each sequence, if it exists, and place it on the same graph as a horizontal asymptote.

3. Find an expression for $b_n$ and one for $a_n$ in terms of $n$.

4. Compute the limit of $\{b_n\}$ as $n \to \infty$ and the limit of $\{a_n\}$ as $n \to \infty$. Compare these with the limits you found in 2.

5. The definition above gives $a_n$ in terms of $b_n$ and $b_{n-1}$. Using this definition, write an expression for $b_n$ in terms of just the $a_i$'s.

6. Use your answers to 4 and 5 to explain in your own words how the sequence $\{a_n\}$ is related to the sequence $\{b_n\}$.

7. Explain in your own words how the limit of $\{b_n\}$ as $n \to \infty$ is related to the sequence $\{a_n\}$.

**Problem**

Repeat the activity, this time starting with the following sequence as $\{b_n\}$:

$$\frac{3}{4}, \frac{6}{6}, \frac{9}{8}, \frac{12}{10}, \frac{15}{12}, \frac{18}{14}, \frac{21}{16}, \ldots$$

## Space station

A space ship is heading towards a space station that is 2.5 miles away. If the space ship travels 1 mile in the next second and then $\frac{1}{2}$ as far each second as in the previous second will it hit the station? If so, when?

### Problems

1. How long will it take the space ship to travel 1.9 miles? How long to travel 1.99 miles? How long will it take until the space ship is .51 miles from the space station?

2. If the space ship travels 3/4 as far in each second as the previous second would it hit the space station? If so, when?

3. If the space ship travels $x$ times as far in each second as the previous second, give an expression involving $x$ that tells how far the space ship travels in five seconds. How far does it travel in $n$ seconds? What values of $x$ would result in the ship hitting the space station? If so, when?

4. Summarize these results in terms of the geometric series

$$1 + x + x^2 + x^3 + x^4 + \cdots .$$

## Decimal of fortune

What follows is a description of a game for two people, Player A and Player B. The object of the game is for Player B to determine a number that has been selected by Player A (and is unknown to Player B as the game begins). A score is computed based on how close Player B has come to Player A's number at the end of the game. Play requires a calculator.

Player A writes down a decimal with eight decimal places. This decimal must be between zero and one. Player B enters the value zero into the calculator. Player B will begin play by selecting a number from List E (evens) or List O (odds). If the number is chosen from List E, it is added to the value in the calculator. If the number is selected from List O, it is subtracted from the value in the calculator. List E is the infinite list of numbers: $1/2, 1/4, 1/6, 1/8, \ldots, 1/2n, \ldots$ List O is the infinite list of numbers: $1, 1/3, 1/5, \ldots, 1/(2n-1), \ldots$ Player A then tells Player B whether or not the value in the calculator is greater than Player A's selected number, less than Player A's selected number, or equal to Player A's selected number. If the value in the calculator is equal to Player A's number, play terminates. If not, Player B selects another number *that has not been selected before*, from either List E or List O and adds it to the value in the calculator if it is from List E or subtracts it from the value in the calculator if it is from List O in an effort to get closer to Player A's number. Play continues until Player B determines Player A's number or until Player B has chosen 20 numbers from the lists. (Player B may not use any number from the lists more than once, but may choose numbers in any order.) In the latter case, Player B may make a final guess (without further calculations on the calculator).

Player B's score is 20 points if Player A's number is determined exactly; otherwise it is $n$ points if the value guessed matches $n$ decimal digits exactly.

Notes: Player B needs to keep track of those numbers that have been used. It is suggested that numbers be written down as they are used while play is in progress. It is also suggested that values in the calculator be written down (or stored in memory) as they appear, so that if a mistake is made entering a number the previous value may be recovered.

For example, a sample game where Player A chooses 0.42000000 (unknown to Player B) is given on the next page.

Play this game four times, with each partner assuming the role of Player A twice. Then answer the following questions.

1. What strategies were developed for Player B as the games were played?

2. What strategies were developed by Player A to prevent Player B from determining Player A's number?

3. If the game were to continue "indefinitely," do you think that Player A's number could be determined exactly? Why or why not?

| Player B chooses | New calculator value | Player A responds |
|---|---|---|
| 1/2 | $0 + 1/2 = .5$ | Too high |
| 1/3 | $.5 - 1/3 = .166\ldots$* | Too low |
| 1/6 | $.166\ldots + 1/6 = .3333\ldots$ | Too low |
| 1/10 | $.3333\ldots + 1/10 = .4333\ldots$ | Too high |
| 1/15 | $.4333\ldots - 1/15 = .3666\ldots$ | Too low |
| 1/30 | $.3666\ldots + 1/30 = .40000$ | Too low |
| 1/60 | $.40000\ldots + 1/60 = .41666\ldots$ | Too low |
| 1/100 | $.41666\ldots + 1/100 = .42666\ldots$ | Too high |

Etc. (What number would you choose next?)

4. Once you get a successive combination of "too high, too low," or vice versa, can you give an upper bound on how far off Player B is?

## Problems

1. Can you determine a number $N$ and a strategy so that Player B can always determine Player A's number (which contains eight decimal places) within $N$ calculator entries?
2. Suppose the rules were changed so that Player B could pick from either list, as before, but that the number that had to be chosen from the selected list was the first unused number. If play were to continue indefinitely, would Player A's number always be determined?
3. Develop an extension of this game by varying one or more of the rules.

*The number 1/3 is subtracted because 1/3 is from List O.

## Approximating functions with polynomials

1. Find a polynomial $p(x)$ with $p(0) = 2$ and $p'(0) = 3$.

2. ... and with $p''(0) = 4$.

3. ... and with $p^{(3)}(0) = -2$.

4. If $p(x)$ is a polynomial, say $p(x) = a_0 + a_1x + a_2x^2 + \cdots + a_nx^n$, find $p^{(k)}(0)$, for $k = 0, 1, 2, \ldots, n+1$.

5. If you had a finite list of numbers, $c_0, c_1, \ldots, c_n$, can you find a polynomial $p(x)$, such that $p^{(k)}(0) = c_k$, $k = 0, 1, \ldots, n$?

### Problem

Find a polynomial $p(x)$ such that $p^{(k)}(0) = k!$, $k = 0, 1, \ldots, n$. (Recall that $k! = k(k-1)\cdots 1$.)

## Introduction to power series

In many cases, it is useful to approximate a function $f(x)$ near some point (say $x_0$) by a polynomial $p(x)$. One example is the tangent line, which is a first-degree polynomial. Recall that the tangent line is the polynomial whose value and derivative both agree with the value and derivative of the function $f$ at $x_0$. Letting $t(x)$ denote the tangent line, we have chosen $t(x)$ to satisfy $t(x_0) = f(x_0)$ and $t'(x_0) = f'(x_0)$. Not surprisingly, by taking a quadratic, it is possible to "match" the function, the first derivative, and the second derivative (at $x_0$). And, in fact, by taking an $n^{\text{th}}$-degree polynomial, we can match the function and the first $n$ derivatives. In this activity we will discover how to do that. To simplify things a bit, for now we will take $x_0 = 0$.

1. Let $p(x) = a_0 + a_1x + a_2x^2 + ...a_nx^n$. Find $p(0), p'(0)$, and $p''(0)$. Find $p'''(0)$. Can you guess $p^{(4)}(0)$? Check this.

2. For $j \leq n$, what is $p^{(j)}(0)$?

3. Suppose $f$ is a function whose value and first 5 derivatives at $x = 0$ are 1. (Can you think of such a function?) What $5^{\text{th}}$-degree polynomial matches $f$ and its first five derivatives at $x = 0$?

### Problems

1. Let $f(x) = \frac{1}{(1-x)}$. Find the first three derivatives of $f$ at $x = 0$. (It is probably easiest if you think of $f(x)$ as $(1-x)^{-1}$.) Guess the fourth derivative, and check that your guess is correct. What $4^{\text{th}}$-degree polynomial approximates $f(x)$ near $x = 0$? What $n^{\text{th}}$- degree polynomial approximates $f(x)$ near $x = 0$?

2. Let $f(x) = \cos x$. Find the first four derivatives of $f$ at $x = 0$. What $4^{\text{th}}$-degree polynomial approximates $f(x)$ near $x = 0$? What $n^{\text{th}}$-degree polynomial approximates $f(x)$ near $x = 0$?

## Graphs of polynomial approximations

In this activity, we will look for a way to approximate a non-polynomial function by a sequence of polynomial functions of higher and higher degree. If we choose the approximating polynomials carefully, each will match the original function more and more closely in the vicinity of a particular chosen point.

Let $f(x) = \dfrac{1}{1+x}$. We will construct approximating polynomials

$$P_0(x), P_1(x), P_2(x), \ldots \text{ where}$$

$P_0(x)$ is a constant function,

$P_1(x)$ is a linear function,

$P_2(x)$ is a quadratic function,

$P_3(x)$ is a cubic function,

$\vdots$

We will focus on the base point $x = 0$.

For this example, we can obtain the polynomials we want by performing a little algebraic "magic."

1. Divide the number 1 by the expression $(1 + x)$, using the long division algorithm. Carry the algorithm out enough to yield at least seven terms of the quotient.

From this result, you can probably guess how the rest of the long division process would go. The quotient is an infinite series in powers of $x$, or a *power series in $x$*.

Now, rather than focusing on the entire quotient, let's take only the first seven terms—which turns out to be a polynomial of degree six. We will use this polynomial for $P_6(x)$.

We started out with the idea of finding polynomials that would serve as good approximations of our original function $f(x) = \dfrac{1}{1+x}$. Let's see how good a job $P_6$ seems to do.

2. Use your graphing calculator or the graphing software on your computer to graph $f$ and $P_6$ on the same set of axes. Graph both functions on the open interval $(-1, 1)$.

3. Does $P_6$ seem to be a good approximation for $f$ near $x = 0$?

4. Describe in words why $P_6$ is or is not a close match for $f$ near $x = 0$.

You will probably agree that it is important for us to decide just what we mean when we say that two functions "match closely" or that one is a "good approximation" for the other near the point $x = 0$.

To do this, we will look at the lower-degree polynomials that our division algorithm yields. That is:

$$\begin{aligned} P_0(x) &= 1 \\ P_1(x) &= 1 - x \\ P_2(x) &= 1 - x + x^2 \\ P_3(x) &= 1 - x + x^2 - x^3 \\ P_4(x) &= 1 - x + x^2 - x^3 + x^4 \\ &\vdots \end{aligned}$$

5. Graph $f$ and $P_0$ on the same set of axes. Graph both functions on the open interval $(-1, 1)$. Use your graphing calculator or the graphing software on your computer.

6. Does $P_0$ appear to be a good approximation to $f$? What do the graphs of $P_0$ and $f$ have in common? Do you think that any reasonable approximating polynomial ought to have this property?

7. Now graph $f$ and $P_1$ on the same set of axes. Graph both functions on the open interval $(-1, 1)$. Use your graphing calculator or the graphing software on your computer.

8. Does $P_1$ appear to be a good approximation to $f$? What do the graphs of $P_1$ and $f$ have in common? Do you think that any reasonable approximating polynomial ought to have this property? Do you believe this is the most appropriate way to approximate $f$ near $x = 0$ *using a linear function*?

Let's look at your last few answers. You probably discovered that $P_0$ and $f$ have the same value at $x = 0$. You probably also decided that it seems reasonable to insist that any good approximation have the same value at the base point, $x = 0$, as the function $f$. Since our first approximating polynomial had to be a constant function, the only possible choice of the constant was $f(0)$. That is:

$$P_0(0) = f(0).$$

9. Verify that $P_1$, $P_2$, $P_3$, $P_4$, $P_5$, and $P_6$ all take on the value $f(0)$, or 1, when $x = 0$.

Now let's look at $P_1$. You undoubtedly identified this linear function correctly as the tangent line to the graph of $f$ at $x = 0$. So the slope of $P_1$ is the derivative of $f$ at $x = 0$, $f'(0)$. Of course, the slope of a line is the (first) derivative of the corresponding linear function. So we see that $P_1'(0) = f'(0)$. The tangent line also passes through the point $(0, f(0))$. So the linear function $P_1$ and the original function $f$ are related by two conditions:

$$P_1(0) = f(0) \text{ and } P_1'(0) = f'(0).$$

We'll continue to examine the approximating polynomials in a similar fashion.

10. Graph $f$ and $P_2$ on the same set of axes. Graph both functions on the open interval $(-1, 1)$.

11. Verify that $P_2$ and $f$ have the same value and the same slope at $x = 0$. I.e., $P_2(0) = f(0)$ and $P_2'(0) = f'(0)$. Describe an additional similarity between $P_2$ and $f$ near $x = 0$.

The additional similarity, of course, has to do with concavity. Both graphs are concave up near $x = 0$. Concavity is determined by a function's second derivative.

12. Verify that $P_2''(0) = f''(0)$.

**Problems**

1. Graph $f$ and $P_3$ on the same set of axes. Graph both functions on the open interval $(-1, 1)$. Verify that $P_3(0) = f(0)$, $P_3'(0) = f'(0)$, $P_3''(0) = f''(0)$, and $P^{(3)}(0) = f^{(3)}(0)$.

2. Compare the graphs of $f$ and $P_4$ near $x = 0$. Find and verify *five* properties that $f$ and $P_4$ share.

3. Generalize your answer to problem 2 to describe the $n^{th}$ approximating polynomial, $P_n$.

## Taylor series

Recall that to construct a power series in $x$ for a function $f$, the coefficient of $x^n$ is

$$a_n = \frac{f^{(n)}(0)}{n!}.$$

It is also possible to construct a power series for $f$ in $x - a$. In this case, the coefficient of $(x - a)^n$ is

$$a_n = \frac{f^{(n)}(a)}{n!},$$

and the resulting series is

$$\sum_{n=0}^{\infty} a_n (x-a)^n.$$

We will call this the power series for $f$ at $a$. As a simple example, we will find the power series in for the function $f(x) = x^3$ at 2.

1. Find the power series for $f(x) = x^3$ at 2. That is, $a = 2$, and we are finding a series in terms of powers of $(x - 2)$.

2. Check your work by multiplying out the $(x - 2)^n$ terms and simplifying.

**Problem**

Find the power series for $f(x) = \ln x$ at 1.

## Approximating logs

1. Show that the Taylor series for $\ln x$ about $x = 1$ is given by

$$(x-1) - \frac{(x-1)^2}{2} + \frac{(x-1)^3}{3} - \frac{(x-1)^4}{4} \pm \cdots \qquad (*)$$

2. Show that $(*)$ converges for $0 < x \leq 2$. In particular, note that the series diverges for $x > 2$.

3. Use the Taylor series to approximate $\ln 1/3$.

4. Use the estimate in 3 to approximate $\ln 3$.

5. How accurate is your estimate of $\ln 3$? (Use a calcalculator to check.)

6. In general, how can one use $(*)$ to estimate $\ln x$ for $x > 2$?

**Problem**

1. Estimate $\ln 5$.

## Using series to find indeterminate limits

You may (or may not) recall the value of $\lim_{x\to 0}\frac{1-\cos x}{x}$. Here we will use power series to find it.

First, recall that the power series for $\cos x$ is $1-\frac{x^2}{2!}+\frac{x^4}{4!}-\cdots$. So,

$$1-\cos x=\frac{x^2}{2!}-\frac{x^4}{4!}+\cdots.$$

Dividing by $x$, we have

$$\frac{1-\cos x}{x}=\frac{x}{2!}-\frac{x^3}{4!}+\cdots;$$

so it follows that

$$\lim_{x\to 0}\frac{1-\cos x}{x}=0.$$

This idea is easy enough: compute the power series for the numerator and denominator, divide, and look at the limit.

1. Find $\lim_{x\to 0}\frac{1-\cos x}{x^2}$.

2. Find $\lim_{x\to 0}\frac{1-\cos x}{x^3}$.

3. Find $\lim_{x\to 0}\frac{\sin x}{x}$.

4. Find $\lim_{x\to 0}\frac{\sqrt{1-\cos x}}{x}$. (Hint: How does this problem compare to 1?) (Note: Try l'Hôpital's rule. Can you figure out a way to obtain this limit using l'Hôpital's rule?)

## Using power series to solve a differential equation

Our goal in this activity is to find a power series ("infinite polynomial") that solves a differential equation. We will work specifically with the differential equation $y' = 2y$ satisfying the initial condition $y_0 = 1$ (which means that $y = 1$ when $x = 0$ or, more simply, $y(0) = 1$). Since you know the solution is given by $y = e^{2x}$, this will also serve as a derivation for the power series for $e^{2x}$. However, nothing we do here depends on our knowing the solution beforehand; indeed, this technique can be used to find a solution to a differential equation that can't be solved in other ways.

Problem: Find a solution to $y' = 2y, \quad y_0 = 1$.

1. Begin with a constant polynomial, $p_0(x) = c_0$. Be sure it satisfies the initial condition.

2. Now construct a polynomial, $p_1(x)$, whose derivative is $2p_0(x)$. Be sure it satisfies the initial condition.

3. Next, construct a polynomial, $p_2(x)$, whose derivative is $2p_1(x)$. Be sure it satisfies the initial condition.

4. Next, construct a polynomial, $p_3(x)$, whose derivative is $2p_2(x)$. Be sure it satisfies the initial condition.

5. What series solves the differential equation $y' = 2y, \quad y_0 = 1$?

## Problems

1. Find a series solution to $y' = 2y, \quad y_0 = 2$. What function does this series represent?

2. Find a series solution to $y' = y, \quad y_0 = 1$. What function does this series represent?

3. Find a series solution to $y' = x + y, \quad y_0 = 1$. What function does this series represent?

## Second derivative test

Suppose all the derivatives of $g$ exist at 0 and that $g$ has a critical point at 0.

1. Write the $n^{th}$ Taylor polynomial for $g$ at 0.

2. What does the second derivative test for local maxima and minima say?

3. Use the Taylor polynomial to explain why the second derivative test works.

### Problem

Suppose in addition that $g''(0) = 0$. What does the Taylor polynomial tell you about whether $g$ has a local maximum or minimum at 0?

## Padé approximation[2]

In this problem, you will find a rational approximation to the exponential function. (Such approximations are called Padé approximations.)

1. Let $f(x) = \dfrac{a+bx}{1+cx}$, where $a$, $b$, and $c$ are constants. Find $f(0)$, $f'(0)$, and $f''(0)$.

2. Let $g(x) = e^x$. Find $g(0)$, $g'(0)$, and $g''(0)$.

3. By equating the values of you found (i.e., making $f$ and $g$ have the same function value, first derivative, and second derivative at 0), find $a$, $b$, and $c$ so that $f(x)$ approximates $e^x$ as closely as possible near $x = 0$.

4. Graph $f(x)$ and $e^x$ on the interval $[-3, 3]$.

[2] Adapted from Deborah Hughes-Hallett, Andrew Gleason, *et al.*, *Calculus*, John Wiley, 1992.

5. Comment on the value of using this method to approximate $e^x$.

## Problems

1. How would you find the Padé approximation for $f(x) = \cos x$?

2. How would you find the Padé approximation for $f(x) = \sin x$?

3. How could you find a more accurate Padé approximation for $e^x$?

## Using Taylor polynomials to approximate integrals

In this activity, you will examine and compare two methods for approximating $\ln 2$ using Taylor polynomials. The methods are useful in estimating the value of many definite integrals where one does not know an antiderivative for the integral or—even if an antiderivative is known—if the needed values of the antiderivative are not easily obtained.

Assume your calculator has a broken $\boxed{\ln x}$ button. That is, assume we need to estimate $\ln 2$ by estimating the value of $\int_1^2 \frac{1}{x}\,dx$, and we only have elementary operations of arithmetic available. We will use two methods.

**Approximate the integral** Let $F(x) = \int_1^x \frac{1}{t}\,dt$. Find the fifth-degree Taylor polynomial for $F$ at 1. Call this polynomial $P_5$. Now, estimate $F(2)$ by computing $P_5(2)$.

**Approximate the integrand** Let $f(t) = 1/t$. Compute the fifth-degree Taylor polynomial for $f$ at 1. Call this polynomial $p_5$. Now estimate $\int_1^2 \frac{1}{t}\,dt$ by computing $\int_1^2 p_5(t)\,dt$. Do this by determining an antiderivative for $p_5$ and using the fundamental theorem of calculus.

After these two estimates have been computed, answer these questions.

1. What is the significance of using a Taylor polynomial at 1?

2. Which of these estimates is more accurate?

3. Describe the difference that results in these two methods.

4. In general, would you expect one of these methods to be a superior method for estimating the value of an integral?

## Problems

1. Compile a list of methods that one can use to estimate $\int_1^2 \frac{1}{t}\,dt$.

2. Let $p_2$ be the second-degree Taylor polynomial at 1 for the function $f(t) = 1/t$. Let $q_2$ be a quadratic polynomial that fits the curve $y = 1/t$ at $(1,1)$, $(3/2, 2/3)$, and $(2, 1/2)$. Are $p_2$ and $q_2$ the same? Which will give you a better approximation of $\int_1^2 \frac{1}{t}\,dt$ when you integrate?

3. We will estimate $\int_1^2 e^{t^2}\,dt$ using a variation of the second method. Let $f(t) = e^t$. Find the fifth-degree Taylor polynomial, $P_a$, at $a$. Now, since $f(t) \approx P_a(t)$ for $t$ close to $a$, $f(t^2) \approx P_a(t^2)$ for $t^2$ close to $a^2$. Decide on a good choice for $a$ and defend your choice. Now, compute $\int_1^2 P_a(t^2)\,dt$. Can you give an estimate for your error? Finally, use the Taylor series for $e^t$ and integrate to give a series that converges to $\int_1^2 e^{t^2}\,dt$.

## Complex power series

A complex number can be thought of as an ordered pair of real numbers $(x, y)$—which we will write as $x + iy$—where $i$ is defined as the square root of $-1$. We call $x$ the *real part* of the complex number $x + iy$, and $y$ is called the *imaginary part* of the complex number. We can add, subtract, multiply, and divide complex numbers by treating $x + iy$ as an algebraic expression with the rule that $i^2 = -1$. For example,

$$(3 - 4i) + (6 + 2i) = (3 + 6) + i(-4 + 2) = 9 - 2i$$

$$(3 - 2i)(5 + i) = 15 - 10i + 3i - 2i^2 = 15 - 7i - 2(-1) = 17 - 7i$$

1. Find the following products:

   (a) $(4 + 2i)(4 + 2i)$
   (b) $(\cos\theta + i\sin\theta)(\cos\theta + i\sin\theta)$
   (c) $(ix)^3$, $\quad x$ real
   (d) $(ix)^4$, $\quad x$ real

2. Write the power series expansion (Maclaurin series) for $e^x$.

3. In the expansion for $e^x$, substitute $x = i\theta$ where $\theta$ is real.

4. Your expansion will involve some terms that include $i$ or $-i$ and other terms that only involve real numbers. Group all the terms that only involve real numbers and then group all the terms that involve $i$ or $-i$.

5. The result of 4 should be two power series. The first power series only involves real numbers and the second power series has $i$ or $-i$ in each term.

   (a) Factor $i$ out of the second power series and write it as $i$ times a power series that only involves real numbers.

   (b) Write the result as

   $$(\text{real power series in powers of } \theta) + i\,(\text{real power series in powers of } \theta).$$

   (c) You should recognize each of the power series in 5b as a well known function of $\theta$. Identify these two functions.

   (d) When you substitute these functions into 5b you obtain a formula for $e^{i\theta}$, called Euler's formula. Do the substitution and write down Euler's formula.

6. Euler's formula can be used to help you remember some trig identities.

   (a) First, using the properties of exponents show that

   $$e^{i\theta}\, e^{i\theta} = e^{i2\theta}.$$

   (b) Next, replace each copy of $e^{i\theta}$ by Euler's formula and multiply these two complex expressions.

   (c) Finally, use Euler's formula to replace $e^{i2\theta}$ by a complex expression.

(d) Complex expressions are equal if and only if their real and imaginary parts are equal. Use this fact to obtain a formula for $\cos(2\theta)$ and another formula for $\sin(2\theta)$.

7. Use the formulas in the previous step to find $\cos^2\theta$ and $\sin^2\theta$ in terms of $\cos(2\theta)$.

8. Use the formulas in the previous step to integrate $\cos^2\theta\,d\theta$ and $\sin^2\theta\,d\theta$.

# Part II

# Projects

## Designing a roller coaster

You have been hired by Two Flags Over Ithaca to help with the design of their new roller coaster.

### Part A.

Each individual has been given a path design for a *straight* stretch (no turns) of a proposed roller coaster. There is a support every 10 feet. A safety rule is that a descent can be no steeper than 80° at any point. In addition each design starts with a 45° incline. (Angles refer to the angle that the path makes with a horizontal line.)

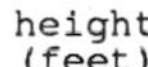


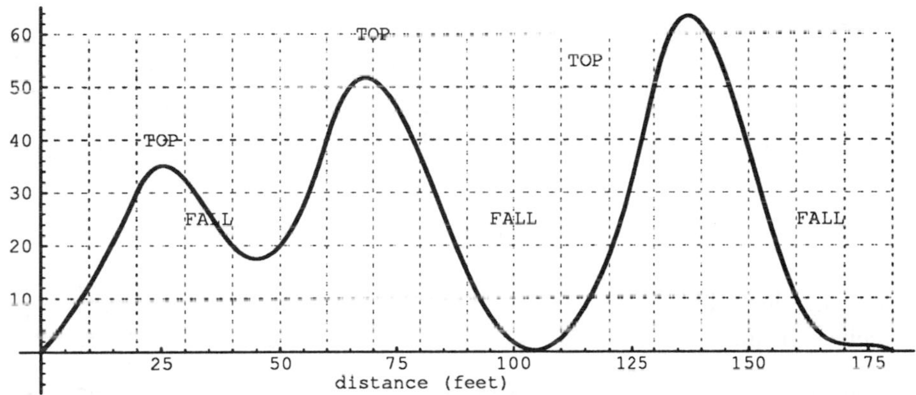


Sample roller coaster

Each individual will report on her/his design. Your report should include the following data.

1. Where is the path increasing and decreasing? (Give your answer in terms of distance along the ground from the start.)
2. Where is the path increasing at an increasing rate, increasing at a decreasing rate, decreasing at an increasing rate, and decreasing at a decreasing rate? (Give your answer in terms of distance along the ground from the start.)
3. For each fall, where is the steepest descent and how steep is the angle at that point?

4. Does your path satisfy the safety criterion? Explain why or why not.
5. Draw the graph of the slope of the path versus distance along the ground from the start.
6. Draw the graph for the rate of change of the slope versus distance along the ground from the start.
7. The *thrill* of the coaster is defined as the sum of the angle of steepest descent in each fall in radians + number of tops. Calculate the thrill of your path.
8. The amount of material needed for a support is the square of the height of the support. So, for example, a support that is 20 feet high would require $20^2$ or 400 feet of material. Find the amount of material needed for the supports in your path.

**Part B.**

Your group should write a report deciding which of the individual paths:

1. is the most thrilling.
2. uses the least material for supports.

See if you can find a rule for finding the place where the slope is steepest in each fall.

**Part C.**

1. Compute how far the coaster would travel along each of the paths your group received in part A.
2. If your coaster must start and finish on the ground and be at least 20 feet high at some point, design the coaster that requires the least amount of support material.
3. Design a path that your group thinks would be the "best" roller coaster if you have 50,000 feet of support material available. Be sure to explain why you think it is the best and any problems with your design. Your report should include how much material is needed for supports, how thrilling the design is, and how far the coaster will travel.

**Part D. (Bonus)**

How fast will the coaster travel?

# Tidal flows

You are a team of consulting engineers studying the flow of water from a certain river into a large lake. You have data in the form of a graph detailing the amount of water that has flowed into the lake from the river over a fourteen-day period. The graph is given in Figure 1.

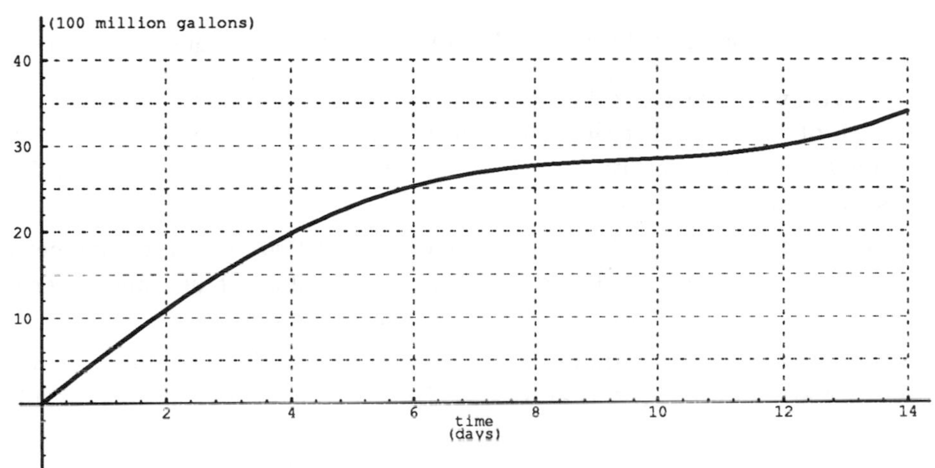


Figure 1

Your job is to report several results to your client as follows:

1. First, determine the total volume of water that has flowed from the river in the first five days of the observation period; in the first ten days; in the entire fourteen-day period.

**Questions 2 and 3 are individual parts.** That is, each group member is to answer the questions for a different day, according to the following list:

Member 1: Day 4

Member 2: Day 6

Member 3: Day 8

Member 4: Day 10

2. Estimate the average rate of flow from three days before your day until your day; from two days before your day until your day; from one day before your day until your day. Also estimate the average rate of flow from your day until three days after your day; from your day until two days after your day; from your day until one day after your day.

3. Estimate the instantaneous rate of flow on your day.

**The remaining questions for this part are to be done by the entire group.**

4. Determine all the time periods when the rate at which the water is flowing is increasing.

5. Determine all the time periods when the rate at which the water is flowing is decreasing.

6. Construct a graph of the rate of flow of the river for the fourteen-day period.

7. Your group's client is considering building a small hydroelectric plant near the mouth of the river. For the project to be feasible, the plant needs to be able to operate at full capacity at least 70% of the time and at reduced capacity at least another 15% of the time.

   To operate at full capacity, a flow rate of at least 200 million gallons per day is required. However, if the flow rate rises above 400 million gallons per day, the equipment is stressed, and the plant must close down.

   If the flow rate is too small for full capacity operation, the plant can still operate at reduced capacity as long as the flow rate remains above 125 million gallons per day.

   Although a statistical analysis is what is needed for a final decision on the plant, what can you tell your client about the two-week observation period that would be relevant to the hydroelectric project?

## Tidal flows

Your engineering group has also been retained to aid in a study of the salinity of a bay near the mouth of a tidal river. A tidal river is one that empties into the ocean or an oceanic bay and is so influenced by the tides that its flow is noticeably affected. That is, the change in water level in the body of water into which the river flows affects the flow of water within the river.

You have collected two kinds of data for a three-day period. The first is a measure of the rate at which the fresh water flows from the river. This can be interpreted as the part of the river's flow rate that is not due to the effect of the tides.

The second is a measure of the effect of the tides. That is, if the river were totally stagnent (not flowing at all), then water would flow into and out of its mouth solely as a result of the tidal rise and fall of the water level in the bay. The second data set provides information about just this aspect of the river's flow rate.

Both data are given in the form of graphs and are shown in Figure 2. Using the given data, provide all the following information.

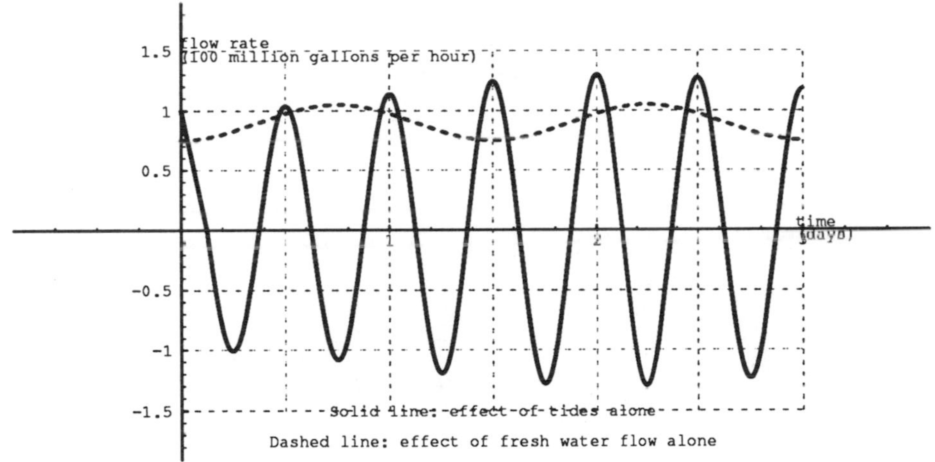


Rate of flow of water from the river into the bay
Figure 2

1. Graph the total net rate of flow of water from the river into the bay.

2. Graph on the same set of axes:

   (a) The total volume of water that would flow from the river from the beginning of the three-day period until time $t$, if there were no tidal influence.

   (b) The total volume of water that would flow from the river from the beginning of the three-day period until time $t$, if the river were stagnent—that is, if the only factor were the ebb and flow of the tides.

(c) The total volume of water that flows from the river from the beginning of the three-day period until time $t$, taking into account both the flow of fresh water and the influence of the tides.

3. When is the water flowing fastest out of the river? Discuss why.

4. Is it ever the case that the water is not flowing at all?

5. What happens when the tide is rising? Explain how this is shown on the given flow graphs and on the total volume graphs that you drew.

## Designing a cruise control

You have been hired as a consultant for a car manufacturer who is designing a cruise control system for a mid-size car. The problem is first broken down into two parts: designing a system to convert real speed into recorded speed; and designing a mechanical apparatus which either slows down or speeds up the car depending upon its recorded speed and a "speed set." We will concentrate on the former problem.

This problem of converting real speed into recorded speed is further broken down into two stages. In the first stage, a metal pin is secured to the inside of the hub of the right front wheel of the car. This pin registers one unit on a counter upon each revolution of the wheel. (See the figure.)

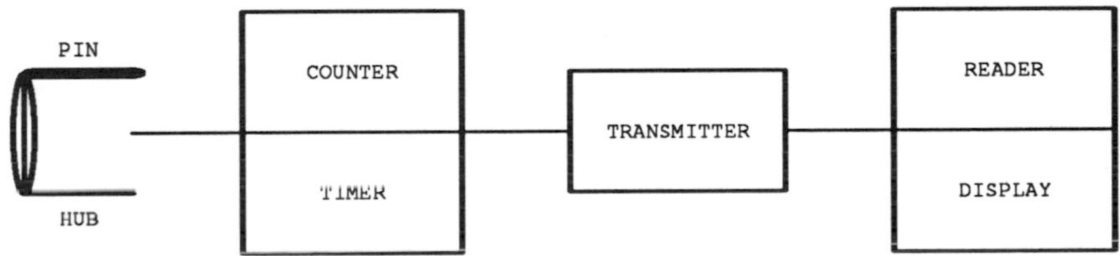


Cruise control

The counter is connected to a timer and a transmitter. In the second stage, the transmitter transmits the total count (i.e., number of revolutions of the wheel) to a "reader" each second and then the counter is cleared (i.e., set back to zero). The reader converts each such count into a recorded speed and displays that speed on a digital speedometer.

The hub of the wheel has a radius of seven inches. Furthermore, the distance from the center of the wheel to the ground in twelve inches.

1. Carefully draw the graph of a function that has as its input (domain) the real speed of the car in miles per hour and has as its output (range) the count that would appear on the counter each second for a car traveling at that speed. Limit the domain of your graph to the interval $[50, 60]$.

2. Carefully draw the graph of a function that has as its input the count per second that could be transmitted by the transmitter and has as its output the recorded speed of the car.

3. Explain how the functions in 1 and 2 should be combined to give a function that has as its input the real speed of the car and as its output the recorded speed of the car. Carefully draw a graph of this function over the interval $[50, 60]$.

4. Assume the cruise control device is completed and installed in a car without any analysis. When the control is "set" at 55 mph there is a problem. Could you have predicted the problem by analyzing the graph in 3 above? Explain.

5. Two design alterations have been suggested to make the system better. One suggestion is to add additional equally-spaced pins around the hub of the wheel. (Say, for example, there would be four pins around the hub.) The formula the reader uses to convert the transmitted count into a recorded speed would have to be changed, of course. The second suggestion would be to transmit the count every 0.5 seconds instead of every second. How would the graphs in 1, 2, and 3 change with either of these suggestions? Would you recommend implementing either or both of these suggestions? Why?

6. Suppose the car manufacturer now wishes to devise an instrument that could keep track of the total distance travelled while the cruise control is activated. How would you suggest that this be done?

## Designing a detector

You are designing a security system for a hospital. The hospital keeps its supply of drugs in a storeroom whose entrance is located in the middle of a 40-foot long hallway. The entrance is a three-foot wide door. The hospital wishes to monitor the entire hallway as well as the storeroom door. You must decide how to program a detector to accomplish this. The detector runs on a track and points a beam of light straight ahead on the opposite wall. The beam reaches from floor to ceiling. Think of the hallway as a coordinate line with the middle of the door at the origin and the hallway to be watched as the interval $[-20, 20]$. You need to determine $x(t)$, where $x(t)$ represents the position of the beam at time t.

The diagram below shows the beam pointing at the origin (i.e., the middle of the door), so if the detector was at this position at some time $t$, we would have $x(t) = 0$. As another example, $x(5) = -15$ means that the beam is pointing at the part of the wall 15 feet to the left of the middle of the door 5 time units after the detector starts.

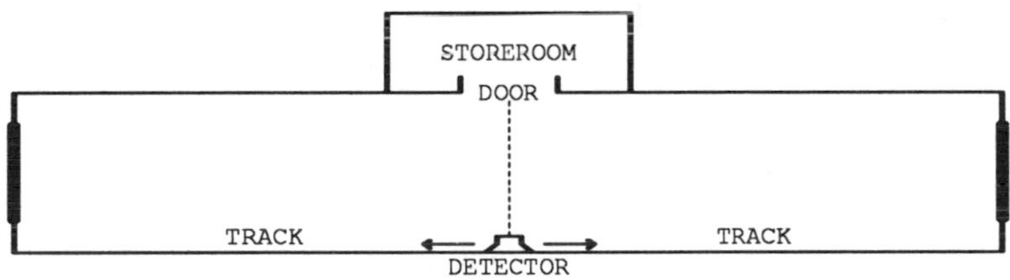


Hallway

### Part A.

1. Draw a graph of $x(t)$ versus time for 10 minutes for what your group thinks is a good choice for $x$. An important part of this is the reasons you think this is a good choice.

2. The beam must stay on an object for at least one tenth of a second in order to detect that object. If the width of a person is one foot, decide whether your answer to 1 will detect a person standing anywhere in the hallway. Explain.

3. Investigate whether an intruder could get to the door by walking down the hallway without being detected by your system. Explain how s/he could do it and how likely you think it is. This may inspire you to to revise your answer to 1.

4. For your answer to 1, compute the longest time that the door will *not* be under surveillance. Remember the door is three feet wide and assume that as long as the beam is hitting any part of the door it is under surveillance.

**Part B.**

1. Find an expression (function) for $x(t)$ for the first 10 minutes. This part of your report should include any restrictions on possible rules for $x(t)$ and reasons for these restrictions. For example, $x(t)$ should never be less than $-20$ because the hall only goes from $-20$ to 20.

2. The beam must stay on an object for at least one tenth of a second in order to detect that object. If the width of a person is one foot, decide whether your answer to 1 will detect a person standing anywhere in the hallway. Explain.

3. Investigate whether an intruder could get to the door by walking down the hallway without being detected by your system. Explain how s/he could do it and how likely you think it is. This may inspire you to to revise your answer to 1.

4. For your answer to 1, compute the longest time interval that the door will *not* be under surveillance.

**Part C.**

Compare parts A and B.

**Part D. (Bonus)**

Answer A.3 and B.3 if the person is running.

# Taxes

Suppose $t(x)$ represents the amount of tax you pay (dollars) if your income is $x$ (dollars). There are (at least) three different ways of depicting this tax on a graph:

- graph $y = t(x)$
- graph $y =$ "average tax rate" ($t_a(x) = t(x)/x$)
- graph $y =$ "marginal tax rate" ($t_m(x) =$ the tax rate on the next dollar $= (t(x+1) - t(x))/\$1$; division by \$1.00 converts dollars to a percentage)

For example, if

$$t(x) = \begin{cases} 0.1x, & 0 \leq x \leq 1000 \\ 0.2x - 100, & x > 1000 \end{cases}$$

then

- tax on \$500 is $0.1 \times \$500 = \$50$
- tax on \$5000 is $0.2 \times \$5000 - \$100 = \$900$
- average tax rate at \$500 is $\$50/\$500 = 10\%$
- average tax rate at \$5000 is $\$900/\$5000 = 18\%$
- marginal tax rate at \$500 is $(\$50.10 - \$50)/\$1 = 10\%$
- marginal tax rate at \$5000 is $(\$900.20 - \$900)/\$1 = 20\%$

## Part A. (Group work)

For the function $t(x)$ above

1. Plot the graphs of $t(x)$, the average tax rate, and the marginal tax rate for $\$0 \leq x \leq \$10{,}000$.
2. Think about plotting each graph for $\$0 \leq x \leq \$10{,}000{,}000$. Describe the behavior of each graph.
3. Give a brief description in English of the function. (If you were writing the tax instruction booklet, this description would be directions for computing the tax.)

### Part B. (Individual work)

1. Take your ID number and turn it into a tax function using the following model. ID # 411-43-7527 gives tax

$$t(x) = \begin{cases} 0.75x, & 0 \leq x \leq 43{,}000 \\ 0.27\,(x - 43{,}000) + 32{,}250, & x > 43{,}000 \end{cases}.$$

2. Plot the graphs of $t(x)$, the average tax rate, and the marginal tax rate for $\$0 \leq x \leq \$100{,}000$.

3. As a political consultant, write two paragraphs. The first should argue why the tax is a good idea, to be supplied to candidates endorsing the tax. The second should argue why such a tax would be a bad idea, to be supplied to candidates opposed to the tax.

### Part C. (Group work)

Attached are three different tax schemes—one given as a tax function, one as an average tax rate, and one as a marginal tax rate.

1. For each tax, compute (or estimate, if necessary) the amount of tax for income of \$5000 and \$50,000.

2. For each tax, plot the other two graphs corresponding to that tax. (I.e., where the tax function is given, plot the average tax rate and the marginal tax rate.)

3. For each tax, comment on whatever aspects strike you as interesting, inappropriate, ...

4. Give formulas for as many of the functions above as possible. (Some will be approximations.)

### Part D. (Group work)

The following table gives the number of tax returns, classified by taxable income.

| Taxable income | Number of returns |
|---:|---:|
| \$0– \$10,000 | 12,000,000 |
| \$10,000– \$20,000 | 16,000,000 |
| \$20,000– \$40,000 | 10,000,000 |
| \$40,000– \$60,000 | 5,000,000 |
| \$60,000– \$80,000 | 2,000,000 |
| \$80,000–\$100,000 | 1,000,000 |
| > \$100,000 | 100,000 |

The tax is given by

$$t(x) = \begin{cases} 0.15x, & \$0 \leq x \leq \$20{,}000 \\ 0.28\,(x - 20{,}000) + \$3000, & \$20{,}000 < x \leq \$100{,}000 \\ 0.33\,(x - 100{,}000) + \$32{,}250, & x > \$100{,}000 \end{cases}.$$

1. Estimate the total tax revenue received from all the returns. Explain your method, and try to give a minimum and maximum for the amount the government could receive.

2. Some economists claim that if marginal taxes were cut by 2% then everyone would increase their taxable income by \$2000. If this assumption were true what would happen to the total tax revenue the government receives. Explain and justify.

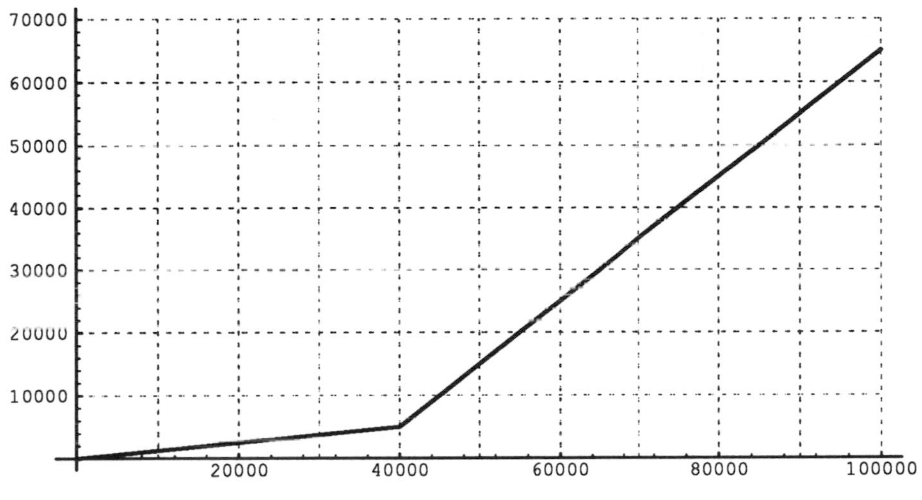


$t(x)$

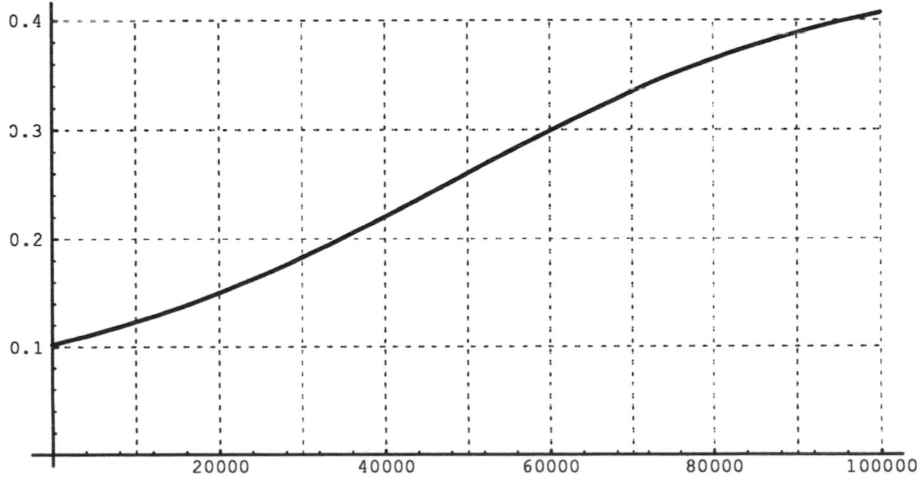


$t_a(x)$

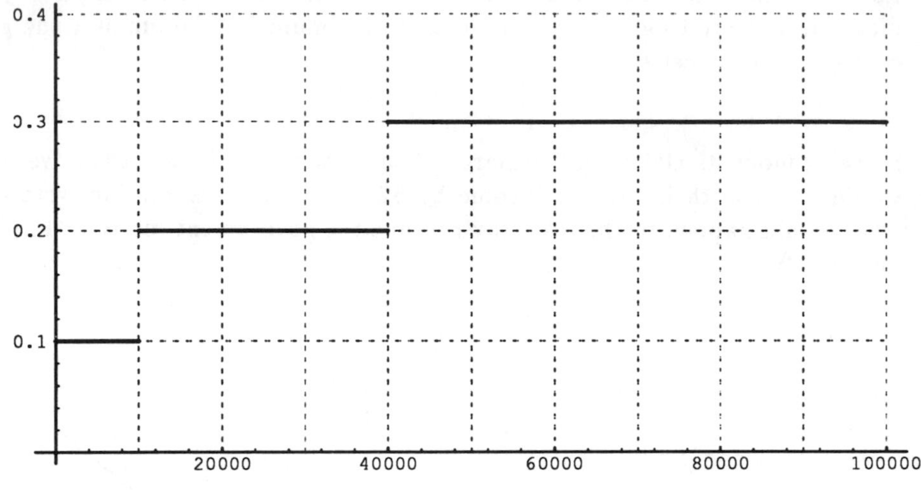


$t_m(x)$

# Water evaporation

## Part A

A water tank was filled with 25,000 gallons of water. Some of the water evaporates each day. The table below gives the amount of water (in thousands of gallons) in the tank $x$ months after the tank was filled for, one and one half months. So for example if $f(1)$ is 21, that would mean that after 1 month there were 21,000 gallons of water in the tank.

1. Using the table of values for the function, each member will graph the function $f$ between 0 and 1.5.
2. Each member of the group will estimate the slope of the graph at the points whose first coordinates are 0, 0.05, 0.10, ..., 1.45, 1.50 (all points from 0 to 1.5 separated by 0.05).
   (a) The first member will use the *forward difference*, which uses the slope of the line segment connecting the point $(x, y)$ and the next point on the table $(x + 0.05, y)$ to estimate the slope at $(x, y)$.
   (b) The second member will use the *backward difference*, which uses the slope of the line segment connecting the point $(x, y)$ and the previous point on the table $(x - 0.05, y)$ to estimate the slope at $(x, y)$.
   (c) The third member will use the *central difference*, which uses the slope of the line segment connecting the previous point $(x - 0.05, y)$ and the next point on the table $(x + 0.05, y)$ to estimate the slope at $(x, y)$.
3. Each member will sketch the slope graph for the given graph.
4. Decide where the *original* function (the function corresponding to the given table) is increasing and concave up, increasing and concave down, decreasing and concave up or decreasing and concave down.
5. Find a relationship between the given graph and the slope graph.

## Part B

Define a new function $g(x) = \dfrac{f(x)}{25}$. This function will satisfy the equation

$$g(kx) = g(x)^k \qquad \text{for all } x \text{ and any number } k. \qquad (*)$$

1. Using the table, verify that $(*)$ holds when $k = 2$ and $x = 0.5$. (You must show that $g(2 \times 0.5) = g(0.5)^2$.)

2. Now verify $(*)$ when $k = 3$ and $x = 0.5$.
3. Use property $(*)$ to find the values of $f(2)$ and $f(4)$.
4. What will happen to $f(x)$ as $x$ gets very large?
5. Sketch the graph of $y = f(x)$ on $[0, \infty)$.
6. Give an example of a function defined on $[0, 1.5]$ that does not satisfy $(*)$. Show at least one specific case where $(*)$ is not true.

### Part C

Decide if the amount of water in the tank will eventually be less than 10,000 gallons. If so how long will it be before this happens?

### Part D

Decide what shape the tank is and what its dimensions are. You may assume that the amount of water that evaporates is proportional to the surface area of the water.

| $x$ | $f(x)$ |
|---|---|
| 0 | 25 |
| 0.05 | 24.75125 |
| 0.1 | 24.50497 |
| 0.15 | 24.26114 |
| 0.2 | 24.01974 |
| 0.25 | 23.78074 |
| 0.3 | 23.54411 |
| 0.35 | 23.30985 |
| 0.4 | 23.07791 |
| 0.45 | 22.84828 |
| 0.5 | 22.62094 |
| 0.55 | 22.39585 |
| 0.6 | 22.17301 |
| 0.65 | 21.95239 |
| 0.7 | 21.73396 |
| 0.75 | 21.5177 |

| $x$ | $f(x)$ |
|---|---|
| 0.8 | 21.30359 |
| 0.85 | 21.09162 |
| 0.9 | 20.88176 |
| 0.95 | 20.67398 |
| 1 | 20.46827 |
| 1.05 | 20.26461 |
| 1.1 | 20.06297 |
| 1.15 | 19.86334 |
| 1.2 | 19.6657 |
| 1.25 | 19.47002 |
| 1.3 | 19.27629 |
| 1.35 | 19.08449 |
| 1.4 | 18.89459 |
| 1.45 | 18.70659 |
| 1.5 | 18.52046 |

# Mutual funds

You are investigating mutual funds for the Securities and Exchange Commission. The Tip-Top-Table fund increased its value by \$50,000 per month for the first 6 months of 1989, and then its value decreased by \$10,000 per month each month after that. The value of the Rags-to-Riches Rule fund changed at the rate of $\$5{,}000t\sin(0.1t^2+1)$ per month $t$ months after the start of 1989. The Go-Go Graph fund changed at the rate per month given on the accompanying graph. The Percentage Growth fund increased 4% per month for the first 10 months of 1989; after that it lost 6% per month. On January 1, 1989, each fund was worth \$1,000,000.

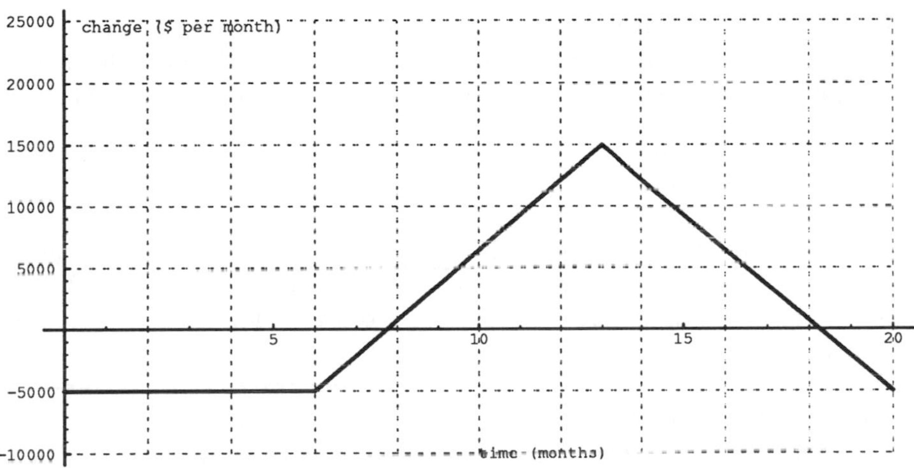


Go-Go Graph fund

1. For *each* fund, give a table of values for the 18 months for each month starting January 1, 1989, of rate of change (in \$/month). (So for $t = 1$ you would need to give the rate of change on February 1, 1989.)

2. For *each* fund, sketch the graph of rate of change versus time for 18 months starting January 1, 1989.

3. Give a formula for the rate of change in terms of time for 18 months starting January 1, 1989, for *each* fund.

4. Answer questions 1, 2, and 3 for the value of funds in terms of time.

5. Assume the funds each continue to perform as they currently are performing. For *each* fund decide which of the following will occur first:

(a) The fund will double in value. (If so, how long will it be until it doubles?)

(b) The fund will be worth less than \$100,000. (If so, when does this happen?)

(c) January 1, 1997.

6. Assume you can transfer amounts from any of these funds to any of the other funds at any time with no commission. If you invested \$100,000 on January 1, 1989, describe how you could end up with the maximum possible value of this investment on January 1, 1991. What is this value?

7. Assume you can transfer amounts from any of these funds to any of the other funds at any time with no commission. If you invested \$100,000 on January 1, 1989, describe how you could end with the least money on January 1, 1991.

8. If the value of your investment in question 7 was less than \$1,000 on January 1, 1991, how fast could you have decreased the value to less than \$1,000? If the value of your investment in question 7 was greater than \$1,000 on January 1, 1991, how long would it take you to decrease the value to less than \$1,000?

9. How could you lose the entire amount as fast as possible? (This is the Brewster's Millions problem.)

An investor invested \$100,000 with a broker on January 1, 1989. The broker invested in the four funds above. His investment is now worth \$50,000. He has written a letter demanding that the broker be prosecuted for incompetence. Decide whether you should prosecute the broker or not. If you decide not to prosecute, write a letter to the complaining investor giving your reasons for not prosecuting. If you decide to prosecute the broker, write the broker a letter explaining why you are prosecuting. In either case, your letter should be backed up with data that will convince the recipient that you have made the correct decision. Your letter should make use of the relevant information found above; include this as an appendix to the letter.

**Bonus.** Answer questions 1, 2, 3, 4, and 5 for the Rapid Rule fund. The Rapid Rule fund's value changes at the rate of $t\sin(0.1t^2+1)$ *per cent* each month.

# Rescuing a satellite

We will investigate whether or not it is possible to rescue an interplanetary probe. The satellite is currently 100,000 miles away from earth. The graph below gives its velocity for the next two years. (After two years the satellite will be useless unless it can be fixed.) The satellite is traveling in a straight line away from earth.

The rescue ship leaves now and its velocity is $v(t) = \sqrt{t + 0.1}$ thousand miles per hour in $t$ years. The rescue ship will also travel in a straight line away from earth, and this is the same as the line on which the satellite is travelling.

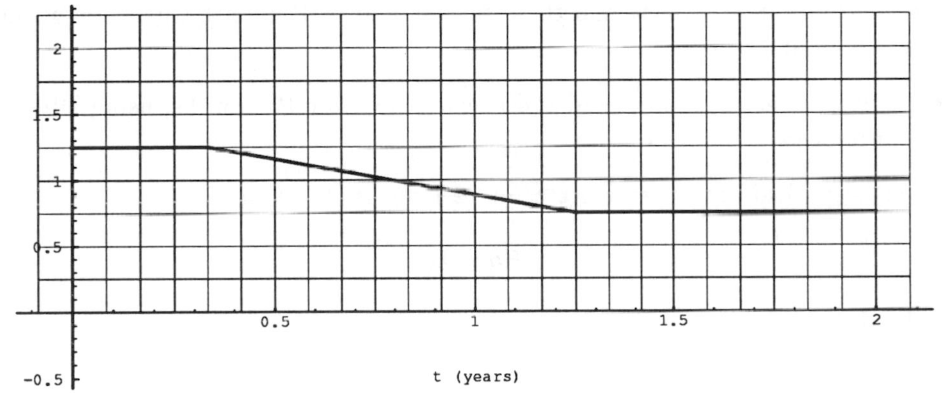


Velocity

Your group must decide whether the rescue ship will catch up with the satellite and if so when.

## Part A

**From the group:** Devise a breakdown, (top-down analysis), of how to solve the problem.

**From each individual:** Use Riemann sums to estimate how far from earth each ship is at certain times. Use $\Delta t = 1$ month.

**First member:**

the rescue ship after 6 months and the satellite after 18 months, using right-hand endpoints;

the rescue ship after 6 months and the satellite after 18 months, using left-hand endpoints;

the actual position of the satellite after 18 months.

**Second member:**

the satellite after 6 months and the rescue ship after 18 months, using right-hand endpoints;

the satellite after 6 months and the rescue ship after 18 months, using left-hand endpoints;

the actual position of the satellite after 6 months.

**Third member:**

the satellite after 12 months and the rescue ship after 12 months, using right-hand endpoints;

the satellite after 12 months and the rescue ship after 12 months, using left-hand endpoints;

the actual position of the satellite after 12 months.

## Part B

Your final report should include:

All the data from part A;

Graphs and rules for the velocities and distances of the satellite and rescue ship, in terms of time.

## Spread of a disease

You are working for the Health Department investigating the spread of a sexually transmitted disease in a population. Statistics are compiled monthly on the numbers of individuals who are infected. Since the disease is spread by contact with an infected person, the number of persons who have the disease at the end of any month depends on the number who had the disease the previous month. Your job is to predict the number of persons with the disease in future months.

Models for the spread of the disease can be thought of in two different ways: (1) as differential equations or (2) as iterations.

If $D(t)$ represents the number of persons with the disease at the end of month $t$, the simplest model (Model 1) has the rate of increase in the number of persons with the disease proportional to the number who have the disease. We express this relationship mathematically with the differential equation $D'(t) = kD(t)$ where $k$ is a constant. This same idea can be expressed as an iteration by $D(n+1) = cD(n)$ where $c$ is a constant.

In these models, $k$ and $c$ are parameters (constants) that can be measured. The Health Department is interested in estimating the parameters and in making long-range predictions of the number of individuals who have contracted the disease. In particular, they wish to know whether, if no cure is found, the numbers reach a steady state equilibrium, or whether the entire adult population is in danger.

**In each group**

The first member should predict the number of persons with the disease for each of the next 10 months and the long-range number if the number of people infected now is 100

1. for $k = 0.3$ (use a differential equation).

2. for $k = -0.7$ (use a differential equation).

3. for $c = 1.3$ (use iteration).

4. for $c = 0.3$ (use iteration).

The second member should predict the number of persons with the disease for each of the next 10 months and the long-range number if the number of people infected now is 100

1. for $k = 0.1$ (use a differential equation).

2. for $k = -0.9$ (use a differential equation).

3. for $c = 1.1$ (use iteration).

4. for $c = 0.1$ (use iteration).

The third member should predict the number of persons with the disease for each of the next 10 months and the long-range number if the number of people infected now is 100

1. for $k = 0.5$ (use a differential equation).

2. for $k = -0.5$ (use a differential equation).

3. for $c = 1.5$ (use iteration).

4. for $c = 0.5$ (use iteration).

A second model (Model 2) assumes that some number of individuals, say $A$, is removed from the population each month. If the disease is fatal, this would be the situation. The the rate at which the disease spreads would be proportional to the diminished infected group. The differential equation $D'(t) = k(D(t) - A)$, and the iteration is $D(n+1) = c(D(n) - A)$.

A third model (Model 3) acknowledges that spread of the disease requires contact between an infected person and a healthy one. In this model, the rate of change is proportional to the product of the number infected and the number not yet infected. Assuming a total population $P$ of adults whose social situation places them in jeopardy, this is expressed by the differential equation $D'(t) = kD(t)(P - D(t))$ or the iteration $D(n+1) = cD(n)(P - D(n))$.

Each person will run the same cases for Model 2 and Model 3 as for Model 1. For Model 2, assume that the value of $A$ is 15; and for Model 3, assume that the value of $P$ is 1000.

### Project requirements

Each individual will submit his/her results for each model. (Each person will be submitting 12 results.) The final group report should include all of the individual results, as well as answers to the following questions.

1. How does the value of $k$ (or $c$) affect the spread of the disease?

2. Does the number of infected individuals reach an equilibrium?

3. Is there any relation between $k$ and $c$?

4. A possible cure for the disease is on the horizon, but will not be generally available until 12 months from now, and then only enough for 500 infected persons can be produced. Will the disease be stoppable at that time, if the cure pans out?

## Tax assessment

A class action suit has been filed against the county board of assessment by a group of landowners unhappy with the assessed value of their properties. Your group will model some assessment schemes and try them out on some lots. Each individual will analyze his/her own lot. Finally your group will analyze one of the disputed lots.

Real estate taxes depend on the assessed value of a property. The value of a piece of property is decided by either a single person (the assessor) or a group (the board of assessment). In most cases, the area of a piece of property is the most important factor in an assessment. However, other factors can give lots with the same area different assessments.

### Part A

For our first example we will look at the simplest case: a rectangular lot that is 200 feet wide and 100 feet deep. This lot is situated in a tax district where land is assessed at $30,000 per acre. (Another way of saying this is to say the tax value of land is $30,000 per acre.)

1. Compute the area of the lot.

2. What is the tax value of the lot?

3. What is the tax value per square foot of the lot?

We will call the *front* of the lot one of the sides that is 200 feet long. If we draw a line parallel to the front through the lot at a distance $x$ feet from the front, then we will call the part of the lot from the front to the line drawn the *first $x$ feet* of the lot. See the picture.

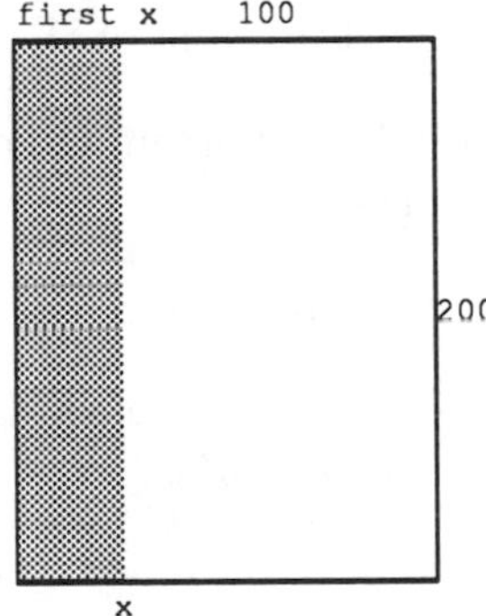


First $x$ feet

4. Find the value of the first $x$ feet in terms of $x$. (We will refer to this as the *first $x$ function*, denoted $f(x)$.)

5. Find the percentage of the value of the entire lot that is due to the first $x$ feet in terms of $x$. (We will refer to this as the *percentage due to the first x feet*, denoted $pf(x)$.)

6. Graph $pf(x)$ for $x$ in $[0, 100]$.

7. What value of $x$ should you chose in order to divide the lot into two lots of *equal tax value*?

### Part B

For many properties, *frontage* is an important attribute of property. For example, the lot pictured has 200 feet of frontage. For commercial property, visibility and accessibility are important, so frontage makes a property more valuable. In such a situation, the front 50 feet of the lot should be worth more than the rear 50 feet of the lot. In the same way the first 25 feet should be worth more than the second 25 feet and so on. A method for finding the value of such a lot is to use what we will call a *worth function*, $w(x)$. The worth function, $w(x)$, is defined as the value of a square foot of land whose center is $x$ feet from the highway.

Answer *all* of the following questions (for a 200 foot by 100 foot lot with 200 feet of frontage) for both of the worth functions:

**(i)** $w(x) = \$(50 - .30x)$ per square foot and
**(ii)** $w(x) = \$(50e^{-.03x})$ per square foot.

1. Compute the area of the lot.

2. What is the tax value of the lot?

3. What is the average tax value per square foot of the lot?

4. Find the value of the first $x$ feet in terms of $x$.

5. Find the percentage of the value of the entire lot that is due to the first $x$ in terms of $x$.

6. Graph $pf(x)$ for $x$ in $[0, 100]$.

7. What value of $x$ should you choose in order to divide the lot into two lots of *equal tax value*?

### Part C (Individual)

Each member of the group should answer questions 1 through 7 of Part B for the worth function $w(x) = 50 - .30x$ for one of the *trapezoidal* lots described below.

**Member 1** Frontage 200 feet, one side is perpendicular to the front with length 100 feet, the rear side is parallel to the front with length 320 feet, the fourth side is a straight line.

**Member 2** Frontage 200 feet, one side is perpendicular to the front with length 100 feet, the rear side is parallel to the front with length 75 feet, the fourth side is a straight line.

**Member 3** Frontage 90 feet, one side is perpendicular to the front with length 100 feet, the rear side is parallel to the front with length 120 feet, the fourth side is a straight line.

**Bonus.** Also answer for $w(x) = \$50e^{-.03x}$ per square foot.

## Part D (Group)

A lot with 200 feet of frontage and 100 feet deep is bounded on one side by a creek that flows through a culvert that crosses under the highway. The distance of this side of the lot from the side perpendicular to the highway is given by $150 + 50\cos(.01\pi x)$ feet when you are $x$ feet from the highway.

1. Sketch the lot.
2. Find the area of the lot.
3. Find the tax value of the lot for the worth function $w(x) = \$(50/(1 + .01x))$ per square foot.

**Bonus.** What value of $x$ should you choose to divide the lot into two parts of equal value?

## Dome support in a sports stadium

The design of a large-scale building requires consideration of aesthetic, societal, and structural factors. The latter include attention to the effects of thermal expansion, "differential" (uneven) settling, dynamic ground motions (earthquakes, etc.), and the "load" on portions of the building resulting from the weight of the building itself on substructures and the wind. In this project, we will be concerned with the construction of a sports stadium. Specifically, we will confine our interest to the support of the dome roof. The county government where the stadium is to be built insists that the roof supports be independent of the seat construction. That is, the buttresses will carry the entire load created by the roof and wind. Furthermore, site considerations require that we support the dome with the aid of vertical buttresses. At least four buttresses must be used. However, the exact number will be your decision. More technical information is given below.

The dome is to be shaped in the form of a spherical cap, made out of homogeneous material, and it must span at least 540 feet. The buttresses must be at least 100 feet high where they touch the roof. They will all be constructed of the same material and will be placed symmetrically around the structure in order that they support equal loads. This material will be cast-in-place concrete. This material was chosen because of its structural properties and the fact that it allows for almost unlimited design possibilities. It is your job to submit a design for the buttresses that is both aesthetic and meets the structural criteria outlined below.

The force that the roof applies to the buttresses as a result of its construction amounts to 5 lbs/ft$^2$ of roof surface area. This will be distributed equally to each buttress. An additional force will be applied to the buttresses as a result of the wind. Meteorological data from the construction location suggests that the additional force that must be considered is a function of the maximum height of the dome according to the graph and table given in Figure 1. The sum of these forces will be applied to each buttress equally and in the direction of the line tangent to the dome roof at the point where the dome meets the buttress. This force can be "resolved" into a vertical component, $V$, and a horizontal component, $H$, (see Figure 2) as follows: if a line segment, $F$, is drawn in the direction of the force whose length is equal to the measure of the force, then the measures of $V$ and $H$ are equal to the lengths of the vertical and horizontal line segments required to form a right triangle.

For simplicity, we will assume that there are only two structural tests that buttress design must pass. The "bearing pressure" must not exceed the limiting values for gravel footers and the "shear stress" cannot exceed the "strength" of the concrete.

The bearing pressure is equal to the total load on the footer of the buttress divided by the area of the base of the buttress. The total load on each footer is equal to the sum of the weight of the buttress and $V$. Concrete weighs 150 lbs/ft$^3$. The bearing pressure cannot exceed 10 tons/ft$^2$ for gravel footers.

The quantity $H$/(minimum horizontal cross-sectional area of the buttress) is known as the *shear stress*; this cannot exceed the concrete strength value 25 pounds per square inch.

Evaluation of your solution will be based upon aesthetic design and verification that your design meets the structural requirements.

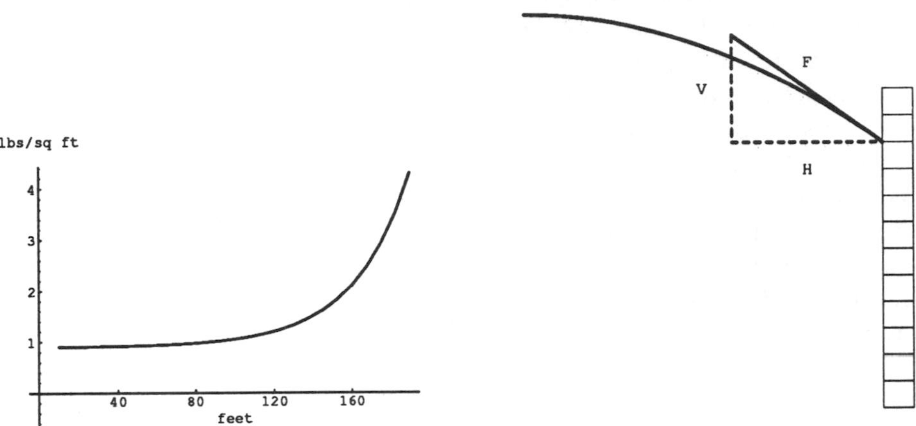


Figure 1

Figure 2

| Height | lbs/sq ft |
|---|---|
| 60 | 0.937 |
| 80 | 0.975 |
| 100 | 1.05 |
| 120 | 1.2 |
| 140 | 1.5 |
| 160 | 2.1 |
| 180 | 3.3 |
| 200 | 5.7 |

## The fish pond

Happy Valley Pond is currently populated by yellow perch. The graph at the right gives an outline of the pond. The pond is fed by two springs: spring A contributes 50 gallons of water per hour during the dry season and 80 gallons of water per hour during the rainy season. Spring B contributes 60 gallons of water per hour during the dry season and 75 gallons of water per hour during the rainy season. During the dry season an average of 110 gallons of water per hour evaporates from the pond, and an average of 90 gallons per hour of water evaporates during the rainy season. There is a small spillover dam at one end of the pond, and any overflow will go over the dam into Bubbling Brook.

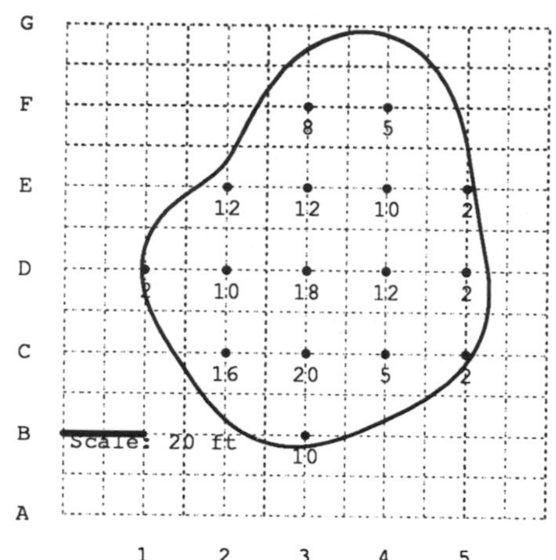


Happy Valley Pond

Spring B has become contaminated with salt and is now 10% salt. (This means that 10% of a gallon of water from Spring B is salt.) The yellow perch will start to die if the concentration of salt in the pond rises to 1%. Assume that the salt will not evaporate but will mix thoroughly with the water in the pond. There was no salt in the pond before the contamination of spring B. Your group has been called on by the Happy Valley Bureau of Fisheries to try to save the perch.

1. Calculate how many gallons of water are in the pond when the water level is exactly even with the top of the spillover dam. Explain how accurate you think your measurement is and why. (A gallon is 231 cubic inches or 231/1728 = 0.13368 cubic feet.)

   The attached table gives a series of measurements of the depth of the pond at the indicated points when the water level is exactly even with the top of the spillover dam.

2. Let $t = 0$ hours correspond to the time when Spring B became contaminated. Assume it is the dry season and that at time $t = 0$ the water level of the pond was exactly even with the top of the spillover dam.

(a) How much salt will be in the pond after $t$ hours? Calculate the amount for $t = 0, 12, 24, 36, 48, 60, 72$.

(b) What is the percentage of salt in the pond after $t$ hours? Calculate the percentage for $t = 0, 12, 24, 36, 48, 60, 72$.

(c) What will the *change* in the amount of salt in the pond be during the time interval $t$ to $t + h$ for some positive $h$.

(d) Write a *differential equation* for the amount of salt in the pond after $t$ hours.

(e) Solve the differential equation and find an expression for the amount of salt in the pond after $t$ hours.

(f) Draw a graph of the amount of salt in the pond versus time for the next three months.

(g) How much salt will there be in the pond in the long run?

(h) Do the fish die? If so when do they start to die?

3. Answer questions 2c–2h if the contamination of Spring B occurred during the rainy season.

4. Answer the questions 2d–2h if the contamination of Spring B occurred during the dry season but after one month the rainy season started.

5. It is very difficult to find where the contamination of Spring B originates so The Happy Valley Bureau of Fisheries proposes to flush the pond by running 100 gallons of pure water per hour through the pond. Your report should include an analysis of this plan and any modifications or improvements that could help save the perch.

Depth of Happy Valley Pond

| Location | Depth | Location | Depth | Location | Depth |
|---|---|---|---|---|---|
| B3 | 10 ft | C2 | 16 ft | C3 | 20 ft |
| C4 | 5 ft | C5 | 2 ft | D1 | 2 ft |
| D2 | 10 ft | D3 | 18 ft | D4 | 12 ft |
| D5 | 2 ft | E2 | 12 ft | E3 | 12 ft |
| E4 | 10 ft | E5 | 2 ft | F3 | 8 ft |
| F4 | 5 ft | | | | |

# Drug dosage

The concentration in the blood resulting from a single dose of a drug normally decreases with time as the drug is eliminated from the body. In order to determine the exact pattern that the decrease follows, experiments are performed in which drug concentrations in the blood are measured at various times after the drug is administered. The data are then checked against a hypothesized function relating drug concentration to time.

Suppose a single dose of a certain drug is administered to a patient at time $t = 0$, and that the blood concentration is measured immediately thereafter, and again after four hours.

(At this point we make a simplifying assumption: namely that when the drug is administered, it is diffused so rapidly throughout the bloodstream that, for all practical purposes, it reaches its fullest concentration instantaneously. Thus, the concentration jumps from a low or zero level to a higher level in zero time. Graphically, this represents a vertical jump on the graph of concentration versus time. [See the graph at the right.] This assumption is probably nearly justified for drugs that are administered intravenously, for example, but not for drugs that are taken orally.)

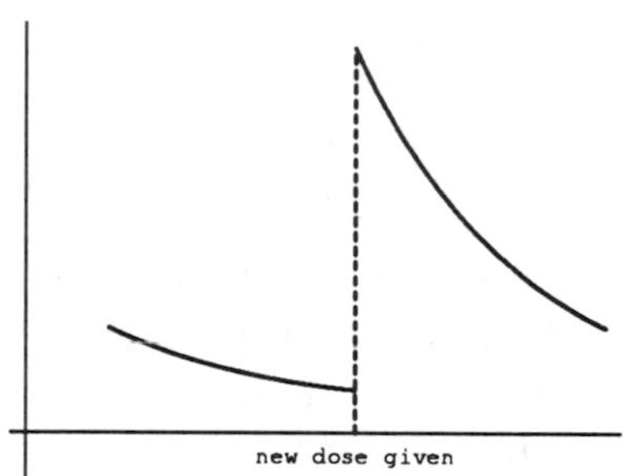


Effect of a drug dose

1. For this part, assume that the function describing concentration as a function of time is linear. There are two data sets given on the last page. Each data set represents a different drug and/or a different initial dose. For each data set:

   (a) Sketch a graph of the concentration function; that is, graph the level of concentration versus time. Assume concentrations are measured in milligrams per milliliter and time is measured in hours.

   (b) Predict the time when the blood becomes free of the drug, assuming no further doses are administered.

   (c) Describe the rate at which the drug is eliminated. Does the rate of elimination seem to depend on any other quantity (e.g., time, level of concentration, etc.)?

(d) Predict what the graph of concentration level versus time would look like if further doses of the drug were administered every six hours for forty-eight hours.

(e) Predict what would happen to the concentration level of the drug if it were administered every six hours indefinitely.

2. Now assume that the rate at which the concentration is decreasing at time $t$ is proportional to the concentration level at time $t$. This idea can be modeled by a *differential equation*, namely, $\frac{dy}{dt} = -ky$, where $y$ is the concentration of the drug in the blood at time $t$, and $k$ is a constant. Using the same data sets as in 1, solve the differential equation, and answer questions 1a to 1e above. (In fact, this model has been shown in clinical tests to be the correct one.)

3. A dilemma facing physicians is the fact that, for most drugs, there is a concentration below which the drug will be ineffective and a concentration above which the drug will be dangerous. See the picture at the right.

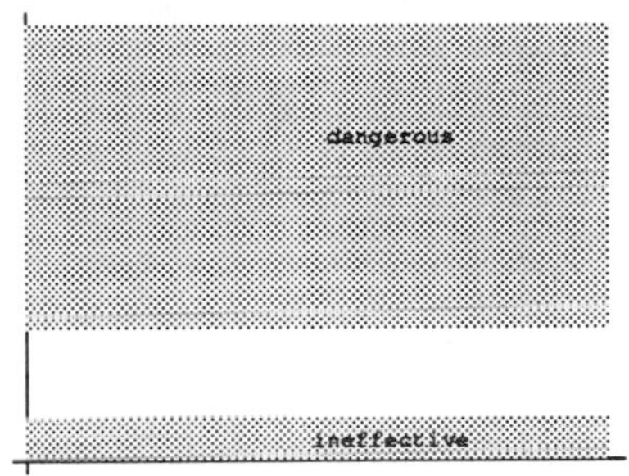


Ineffective and unsafe levels

(a) Suppose that for the drug in experiment 2, the minimum effective level is 0.45 mg per ml and the maximum safe level is 2.15 mg per ml. Are the dose and interval between doses given in the experiment satisfactory for maintaining appropriate concentrations indefinitely? Explain. If the answer is no, can you achieve a satisfactory long-run level just by adjusting the time between doses? Just by adjusting the dose? (Assume there is a simple way to tell just how much substance must be administered in order to raise the concentration by any given amount. That is, you can answer this question by specifying how much the concentration of drug in the blood needs to be raised by each dose.)

(b) Answer the same questions assuming the minimum effective level is 0.5 mg per ml, and the maximum safe level is 1.65 mg per ml.

4. A patient has been taking the drug XL37 regularly for several months. He has experienced some health problems that he attributes to the drug. The maximum safe level of XL37 in the blood stream is known to be 2.0 mg per ml. The patient has been tested periodically for concentration of XL37 in his blood, and the level found has been consistently between 1.9 and 1.95 mg per ml. The patient wishes to initiate a malpractice suit against the clinic, claiming that he has been administered an overdose of the drug.

(a) Taking the point of view of the lawyer defending the clinic, argue why the patient's claim is not correct.

(b) Taking the point of view of the attorney for the patient, argue why there is evidence of an overdose.

(c) As judge in the case, what further information would you like to see before deciding this case?

Data

| | Experiment 1 | Experiment 2 |
|---|---|---|
| Concentration at time $t = 0$ | 1.0 mg/ml | 1.5 mg/ml |
| Concentration after 4 hours | 0.15 mg/ml | 0.75 mg/ml |

## Investigating series

In this project, you will experiment with some infinite sequences and their limits.

### Part A

Starting with a given sequence of numbers, $\{b_1, b_2, \ldots\}$, you will construct a new sequence $\{a_1, a_2, \ldots\}$ as follows:

$$\begin{aligned} a_1 &= b_1 \\ a_2 &= b_2 - b_1 \\ a_3 &= b_3 - b_2 \\ &\vdots \\ a_n &= b_n - b_{n-1} \\ &\vdots \end{aligned}$$

Starting with each of the following sequences as $\{b_n\}$:

(I) $\frac{2}{1}, \frac{8}{3}, \frac{26}{9}, \frac{80}{27}, \frac{242}{81}, \frac{728}{243}, \frac{2186}{729}, \ldots$ and

(II) $\frac{3}{4}, \frac{6}{6}, \frac{9}{8}, \frac{12}{10}, \frac{15}{12}, \frac{18}{14}, \frac{21}{16}, \ldots$

1. Compute the first six elements of the sequence $\{a_n\}$.
2. Graph $\{a_n\}$ versus $n$ and $\{b_n\}$ versus $n$ on the same set of coordinate axes. Plot at least the first six values for each sequence. Visually determine the limit of each sequence, if it exists, and place it on the same graph as a horizontal asymptote.
3. Find an expression for $b_n$ and one for $a_n$ in terms of $n$.
4. Compute the limit of $\{b_n\}$ as $n \to \infty$, and the limit of $\{a_n\}$ as $n \to \infty$. Compare these with the limits you found in 1.
5. The definition above gives $a_n$ in terms of $b_n$ and $b_{n-1}$. Using this definition, write $b_n$ in terms of just the $a_i$'s.
6. Use your answer to 4 to explain in your own words how the sequence $\{a_n\}$ is related to the sequence $\{b_n\}$.

7. Explain in your own words how the limit of $\{b_n\}$ as $n \to \infty$ is related to the sequence $\{a_n\}$.

**Note.** You will be answering seven questions for each of the given sequences. Your answers for sequence (II) will be part of the group report; your answers for sequence (I) are to be written up and handed in individually.

### Part B

Now let $\{a_n\}$ be the sequence $\{\frac{1}{n+1}\}$. Define a new sequence $\{s_n\}$ by $s_n = a_1 + a_2 + a_3 + \ldots + a_n$.

1. For each of $n = 1$ to $n = 8$, compute the value of $s_n$. Now look at $s_1, s_2, s_4, s_8$, and the values of $s_n$ in the table below. How does the value of $s_n$ change when $n$ is doubled?

| n | $s_n$ |
|---|---|
| 1 | 0.5 |
| 2 | 0.833333 |
| 4 | 1.283333 |
| 8 | 1.828968 |
| 16 | 2.439553 |
| 32 | 3.088798 |
| 64 | 3.759276 |
| 128 | 4.440899 |
| 256 | 5.128236 |
| 512 | 5.818466 |
| 1024 | 6.510151 |

2. Draw the graphs of $\ln(x + 1)$ and $s_n$ on the same set of axes. What do you see? Can you explain it?

3. On a second set of axes, draw a picture of the function $y = \frac{1}{x+1}$. Find a way to represent the value of $\ln(n + 1) - \ln 2$ in this picture for $n = 8$.

4. Again, on the same (second) graph, find a way to represent the value of $s_n$. Is there any relationship between $s_n$ and $\ln(n + 1) - \ln 2$? If so what is the relationship?

5. What is the limit of $\{s_n\}$ as $n \to \infty$? Why do you think so?

### Part C

Again, let $\{a_n\}$ be the sequence $\{\frac{1}{n+1}\}$. Define a new sequence $\{t_n\}$ by: $t_n = a_1 - a_2 + a_3 - a_4 + \ldots \pm a_n$.

1. Compute $t_n$ for $n = 2, 4, 6, 8, 10$. Is $\{t_{2n}\}$ an increasing or a decreasing sequence, or neither? Explain.

2. Compute $t_n$ for $n = 1, 3, 5, 7, 9$. Is $\{t_{2n+1}\}$ an increasing or a decreasing sequence, or neither? Explain.

3. What relationship, if any, exists between $\{t_{2n}\}$ and $\{t_{2n+1}\}$? Explain.

4. Assume $\{t_n\}$ converges and has limit $L$. What relationship exists between $\{t_{2n}\}$, $\{t_{2n+1}\}$, and $L$? Explain. Use this relationship to obtain an error bound for $|L - t_{20}|$. Explain your answer.

**Bonus.** Show that $\{t_n\}$ converges to $\ln(2) - 1$. (Hint: Expand $\ln(x+1)$ using a Taylor series about $a = 0$.)

Determine $n$ such that $|t_n - (\ln(2) - 1)| < 0.01$.

Calculus 2 Project 4

## Topographical maps

Attached is a section of a topographical map of Tompkins County (the county in which Ithaca College is situated). The light lines on the map are lines of constant altitude: that is, if you travelled along one of these lines, your altitude would remain constant. In mathematics, we call such lines *level curves.* With some thought and practice, one can picture a three-dimensional surface (in this case, the surface of the earth) by looking at a two-dimensional map on which the level curves have been indicated.

1. One point on your map is labelled "O" (for origin). From that origin, a direction is indicated by an arrow, and dotted lines indicate a field of view.

   Imagine that you are standing at the origin, looking in the general direction of the arrow. Sketch the land forms that you would see from that viewpoint, as they might appear in a photograph of the region. Include the entire field of view in your sketch. (You may find it easier to draw the picture assuming you are hovering several yards above the origin.)

2. Several points on the map have been labelled: A1, A2, B1, B2, etc. The first person in your group will use the "A" points for this part, the second the "B" points, and so forth. For each of your points:

   (a) Imagine that you are situated at the indicated point, moving due north in such a way that your northerly progress is 100 feet per minute. At approximately what rate is your altitude changing? Use positive values to indicate increasing altitude and negative values for decreasing altitude.

   (b) Do the same, assuming that you are moving due east rather than north.

   (c) In what direction should you travel so that your altitude is increasing as rapidly as possible? Indicate the direction in degrees measured clockwise from due north.

3. Imagine that you are a novelist. You have described a landscape for your readers. The landscape includes two mountains, one larger and higher than the other. Each mountain has some foothills. The smaller mountain has one very shear face. Between the mountains is a valley, through which a river flows.

   To make your story more realistic (it involves an adventurer finding her way through the region), you decide to include a topographical map of the scene in the book. Construct this map.

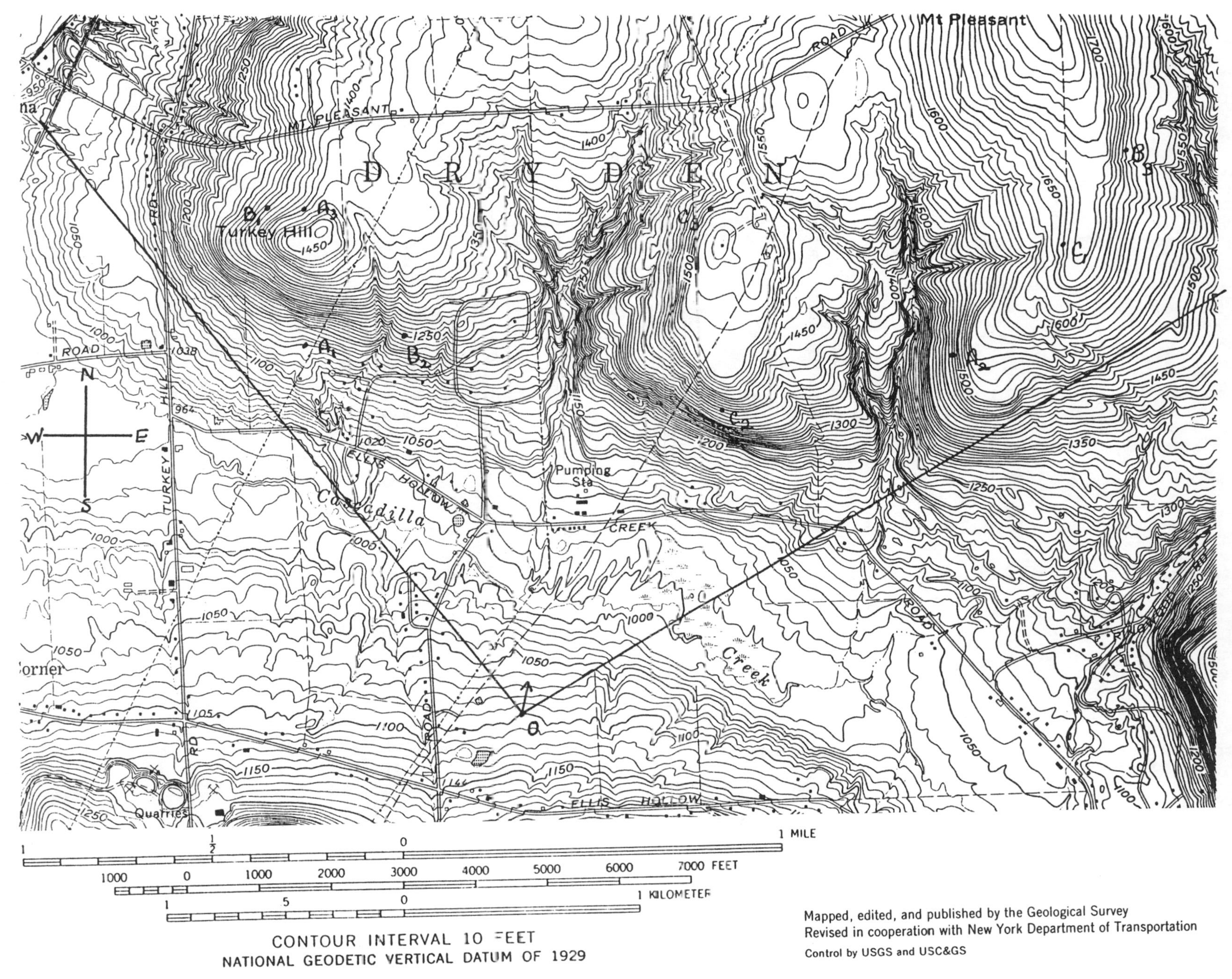
Mt Pleasant
ROAD
MT PLEASANT
DRYDEN
Turkey Hill
1450
1400
1550
1600
1650
1700
1500
1250
1200
1100
1050
1000
1038
964
1020
1300
1350
1150
ROAD
TURKEY
HILL
RD
N
W
E
S
ELLIS
HOLLOW
Cascadilla
CREEK
Pumping Sta
Creek
Quarries
1105
1150
1100
ELLIS HOLLOW
ROAD
orner
1 MILE
7000 FEET
1 KILOMETER
CONTOUR INTERVAL 10 FEET
NATIONAL GEODETIC VERTICAL DATUM OF 1929
Mapped, edited, and published by the Geological Survey
Revised in cooperation with New York Department of Transportation
Control by USGS and USC&GS

# NOTES

## NOTES

# NOTES

# NOTES

# NOTES

# NOTES

# NOTES

# NOTES

# NOTES